GLOBAL AGRICULTURE: DEVELOPMENTS, ISSUES, AND RESEARCH

# GLOBAL AGRICULTURE

# DEVELOPMENTS, ISSUES, AND RESEARCH

# VOLUME 3

# GLOBAL AGRICULTURE: DEVELOPMENTS, ISSUES, AND RESEARCH

Additional books in this series can be found on Nova's website under the Series tab.

Additional E-books in this series can be found on Nova's website under the E-book tab.

GLOBAL AGRICULTURE: DEVELOPMENTS, ISSUES, AND RESEARCH

# GLOBAL AGRICULTURE

# DEVELOPMENTS, ISSUES, AND RESEARCH

## VOLUME 3

MARVIN R. ROBERTSON
AND
JEFFERY I. SCHMIDT
EDITORS

nova publishers
*New York*

For permission to use material from this book please contact us:
Telephone 631-231-7269; Fax 631-231-8175
Web Site: http://www.novapublishers.com



**Library of Congress Cataloging-in-Publication Data**

ISSN: 2168-0434

ISBN: 978-1-62618-260-8

*Published by Nova Science Publishers, Inc. †New York*

# CONTENTS

# PREFACE

This book explores the developments, issues and research outlook of global agriculture. Topics discussed include classifying and measuring agricultural support and identifying differences between the WTO and OECD systems; China's volatile pork industry; long-term growth prospects for wheat production in Afghanistan; and economic policy and cotton production in Uzbekistan.

Chapter 1 – Most countries provide some level of support to their agricultural sectors. Because support can affect producers and consumers in other countries, a number of systems have been developed to measure agricultural support levels and classify types of support in ways that facilitate comparing them across countries. The WTO and OECD employ similar classification systems, generally addressing the same question and measuring the same programs. However, results can be surprisingly and fundamentally different, rendering comparisons inappropriate, meaningless, or even wrong. Careful attention to the sources of difference can prevent potential misunderstandings and misleading uses.

Chapter 2 – With China's emergence as a new source of potential demand for U.S. pork exports, it is important for U.S. farmers, business leaders, and policymakers to understand the volatile nature of China's pork industry. Prices, hog inventories, and pork output in China fluctuate from year to year in response to various factors that influence the market, and China's imports of pork tend to rise when Chinese hog prices are high. Extensive policy intervention by the Chinese government has contributed to consolidation in the country's pork industry but has not stabilized the market. Imported pork is becoming more competitive in China as Chinese pork production costs rise and animal disease outbreaks, environmental threats, and food safety concerns constrain growth of China's hog industry.

Chapter 3 – Afghanistan has had success at achieving higher output for wheat, its key staple food and major crop. From 1990-2009, wheat production grew 3.5 percent per year, driven primarily by yield increases. However, given expected increases in demand, imports are likely to grow in coming years even if this production growth is sustained, suggesting growing dependence on supplies from Pakistan and other countries. For Afghanistan to achieve self-sufficiency in the next decade, more widespread use of improved seed and fertilizer on irrigated and rainfed wheat fields would have to be combined with sharp increases in irrigated wheat area—an unlikely outcome.

Chapter 4 – Uzbekistan is the seventh largest global cotton producer and third largest cotton supplier for world markets. Uzbekistan's Government policies largely shield cotton producers from world market price signals, and cotton area has changed little over the past decade despite strong international price fluctuations. Government pricing and exchange rate policies tax cotton producers and more than offset the value of input subsidies for cotton growers. The degree of taxation declined for several years after 2000, but increased again in the late 2000s. In the 2009, cotton output dropped as Uzbekistan responded to reduced water availability and increased global food prices with higher taxes on cotton growers. With continued taxation of cotton production, Uzbekistan likely will continue to lose ground to more dynamic cotton exporters like India and Brazil. As a result, USDA's longrun baseline projections for Central Asia show that the region's share of world cotton production will continue to fall over the next decade.

In: Global Agriculture ISBN: 978-1-62618-260-8
Editors: M. R. Robertson, J. I. Schmidt © 2013 Nova Science Publishers, Inc.

*Chapter 1*

# CLASSIFYING AND MEASURING AGRICULTURAL SUPPORT: IDENTIFYING DIFFERENCES BETWEEN THE WTO AND OECD SYSTEMS*

*Anne Effland*

## ABSTRACT

Most countries provide some level of support to their agricultural sectors. Because support can affect producers and consumers in other countries, a number of systems have been developed to measure agricultural support levels and classify types of support in ways that facilitate comparing them across countries.

The WTO and OECD employ similar classification systems, generally addressing the same question and measuring the same programs.

However, results can be surprisingly and fundamentally different, rendering comparisons inappropriate, meaningless, or even wrong. Careful attention to the sources of difference can prevent potential misunderstandings and misleading uses.

---

* This is an edited, reformatted and augmented version of United States Department of Agriculture, Economic Information Bullentin, Publication No. 74, dated March 2011.

## SUMMARY

### What Is the Issue?

Most nations provide some level of support to their agricultural sectors. Different types of support can affect producers and consumers both in the supporting country and in other countries. As such, measures of domestic agricultural support are highly contested in the negotiation of trade agreements. Two key systems have emerged for classifying and comparing agricultural support levels across countries. The World Trade Organization's (WTO) notification system produces the Aggregate Measurement of Support (AMS); member governments, in adherence to a formal trade agreement, submit their own data, though such notifications may be submitted irregularly due to lack of capacity or lack of timely and complete data. The Organization for Economic Cooperation and Development's (OECD) Producer Support Estimate (PSE) is also a consensus framework among member nations, but its purpose is to facilitate dialogue on policy reform and effective policy design. The OECD measure relies on data provided by members, supplemented with other sources by OECD experts in order to make annual estimates of transfers across sectors of the economy.

Governments, nongovernmental organizations, researchers, and journalists use both measures regularly to compare the levels and types of support to agriculture across countries. Since both systems produce measures based on the same support programs, some users may attempt to use the OECD measure as a proxy for the WTO measure, which is more narrowly focused and more irregularly reported. But because these two systems were developed for different purposes, they are not identical in their classification schemes, their policy inclusiveness, and their methodologies. These differences can result in surprisingly different results.

### What Did the Study Find?

- The WTO classification system requires members to categorize their programs according to rules regarding their expected trade-distorting impacts, which determines whether those programs are subject to each member's maximum support commitments under the Uruguay Round Agreement on Agriculture. The OECD system classifies programs based on criteria related to program implementation, rather than expected impacts, and programs are separated based on whether support is to producers, consumers, or the agriculture sector generally.

- For the United States, these classification and measurement differences are reflected in a higher level of domestic agricultural support reported under the OECD system compared to the WTO system. From 1995 to 2007, annual domestic support reported under the WTO system ranged from 68 percent to 90 percent of that reported under the OECD system.
- In some cases the two systems employ different methods to measure the same type of support; for example, the methods used by the WTO and OECD systems to calculate market price support (MPS) yield strikingly different results. Because the OECD method uses the gap between two current (domestic and world) prices to calculate MPS, the amount of MPS may vary widely from year to year. When world prices are high, the gap between a supported domestic price and world price will likely be small, reducing MPS; when world prices fall, that gap will likely increase and MPS will be higher. In contrast, the MPS calculated under the WTO system compares the same fixed world reference price (the 1986-88 average) with a domestic administered price, so when the domestic administered price is stable, the WTO's MPS method will result in only slight variation from year to year based on changes in eligible production.
- For the United States, the difference in methodology for calculating MPS results in reported annual support differences ranging from \$3 billion to \$16 billion over 1995-2007. Combined with significantly different methods for classifying direct support to producers, these MPS results contribute to the OECD producer support estimate (PSE) ranging from \$13 billion to \$40 billion higher than the WTO aggregate measurement of support (AMS) over the same period.
- It may be possible to translate from one system to the other, perhaps to recreate a missing year of data or to develop new composite indicators, but the task requires a detailed knowledge of the methodologies used by both systems, a detailed understanding of country policies, sufficient reporting transparency to identify individual programs, and some choices about how to recalculate unique measures, like MPS.

## How Was the Study Conducted?

A comparative framework for analyzing the two domestic support measurement systems was built by examining the origins, purposes, and classification schemes of both through published documentation and their use in reports and databases since the mid-1990s. This framework allows for juxtaposing the detailed classification and measurement methods of each system and making direct comparisons of how they would be applied across a set of country policies. The impacts that the differing categorization and measurement methods could have on domestic support reporting are demonstrated by analyzing their application to U.S. programs and data reporting from 1995 to 2007. The U.S. examples also provide an opportunity to clarify some common misconceptions about comparability between the WTO and OECD systems.

# INTRODUCTION

Most nations provide some level of support to their agricultural sectors. However, that support can take many forms. Different types of support can affect producers and consumers both in the supporting country and in other countries. As such, measures of domestic agricultural support are highly contested in the negotiation of trade agreements. The desire for discipline of levels of support has led to intense efforts to characterize the types of support and to measure their likely impacts on production, trade, and the well-being of producers and consumers. Out of these efforts have come a number of systems to measure agricultural support levels and classify types of support in ways that facilitate comparing them across countries.

Two key systems for classifying and measuring domestic agricultural support have become widely accepted. One is the World Trade Organization's (WTO) Domestic Agricultural Support Notification System, which produces a measure called the Aggregate Measurement of Support (AMS). The other is the Organization for Economic Cooperation and Development's (OECD) Total Support Estimate, which produces a measure called the Producer Support Estimate (PSE) (table 1).

The WTO notification system and AMS exist for the purpose of securing commitments within the legal framework of a formal trade agreement. The rules governing domestic support reporting reinforce the agreement by providing a means through which adherence to commitments can be assessed.

**Table 1. Comparison of the WTO and OECD domestic support measurement system**

| | WTO | OECD |
|---|---|---|
| Purpose | Evaluate<br><br>• Observance of trade agreements | Facilitate<br><br>• Policy dialogue |
| Method | Measure<br>• Trade-distorting support<br>• Based on negotiated criteria related to expected trade impacts | Measure<br>• Support to sector<br>• Based on criteria related to program implementation |
| Outcomes | Annual notifications<br>• Assess adherence to maximum support commitments | Comparative monitoring and evaluation report<br>• Assess policy reform progress<br>• Research database |

Source: Economic Research Service, USDA, based on WTO *Notifications Handbook* and OECD *PSE Manual.*

The OECD classification system and total support estimates are also a consensus framework among members, but their purpose is to facilitate dialogue on policy reform. OECD measures focus on identifying transfers across sectors of the economy and provide a foundation for economic analysis of effective policy design.

The WTO system was developed during the Uruguay Round of negotiations of the General Agreement on Tariffs and Trade (GATT), which brought the world agricultural sector under the umbrella of this longstanding global trade agreement (WTO, 1994). The Uruguay Round Agreement on Agriculture (URAA) became part of the system administered by the World Trade Organization (WTO), which in 1995 superseded the GATT as the institution charged with establishing the rules of trade and providing a forum for members to monitor the world's foremost multilateral trade agreement. Under the URAA, member countries (currently numbering 153) agreed to limit the most trade-distorting types of support provided to their domestic agricultural sectors.

The URAA establishes a legally binding framework and requires annual reporting (notification) of domestic support to ensure transparency among members, all of whom can question each other's notifications at regular meetings of the WTO Committee on Agriculture. WTO domestic support notifications are publicly available, and key summary data are available in spreadsheet format (WTO, *DS:1 and Relevant Supporting Tables*). However, more detailed data are available only through official documents accessed through a system that can be difficult and cumbersome to use.

The agricultural support measure adopted by the OECD—based on earlier work by Josling (1973) and originally called the Producer Subsidy Equivalent—was used in support of the Uruguay Round negotiations to quantify and categorize different policy instruments. In years since, OECD's system has been redesigned more than once and has been renamed the Producer Support Estimate (PSE). As member countries have adopted policies less directly linked to current production of individual commodities, the OECD measure has moved beyond a focus on commodity support and its cost to consumers to a more complex expression of the ways in which governments support their agricultural producers, consumers, and infrastructure.

The current OECD system no longer estimates all support in terms of individual commodity outputs. Initiated in response to the introduction of a more complex array of program designs by member governments, it now captures policy support to producers based on current production factors and farm revenue, historical production, and non-commodity (for example, environmental) outputs. The OECD system supports an annual monitoring and evaluation exercise to assess the progress of agricultural policy reform in member countries, as well as agricultural policy development in selected emerging economies. The system helps members and observers to evaluate policy tools objectively, measure the support provided by their own policies in comparison with others, and develop "best practices" guidance based on empirical evidence. OECD also maintains an annually updated multiyear database accessible to the public (OECD, *PSE/CSE database*).

## How Do the Two Systems Measure up?

The two systems address the same questions—how much support do individual countries provide to their agricultural sectors and what forms does that support take? But while the OECD monitoring and evaluation reports are released on a regular annual schedule, offering users a constantly updated source of data, the WTO domestic support notifications are made available as they are submitted by member governments. The OECD measure relies primarily on data provided by member countries; when the data are incomplete, OECD experts identify other sources or develop estimates in order to meet the requirements of an annual publication. WTO notifications, in contrast, are completed by member governments themselves and may not be submitted regularly for a variety of reasons, including lack of capacity or lack of timely and complete data. Because the regularity of reporting varies widely, WTO notifications are a less dependable source of data from which to draw comparable measures of domestic support across countries.

Both the OECD and WTO measures have been used by journalists, governments, non-governmental organizations, and researchers to report on the annual levels of support to agriculture, to provide a basis for comparing domestic support across countries in the context of trade negotiations and domestic policymaking, and to assess the character and potential impacts of various types of domestic support on markets and trade. Economists in particular have made use of these measurement systems in both qualitative and quantitative analyses of trade policy reform (Burfisher, 2001; Takahashi, 2009; Orden et al., 2011).

Since both systems measure essentially the same support programs, it may seem reasonable to use the OECD measure (for those countries for which PSEs are available[1]) as a proxy for the WTO measure to account for those years in which WTO notifications data are not available for all members. But because these two measures were developed for different purposes, they are not identical in their classification schemes, their policy inclusiveness, and their measurement methodologies.

These differences can result in surprisingly different support totals, as a comparison of the WTO and OECD support totals for the United States shows (figure 1). From 1995 to 2007, annual domestic support reported under the WTO system ranged from 68 percent to 90 percent of that reported under the OECD system.

## WTO Domestic Agricultural Support Notifications

The WTO system provides for annual reporting of domestic support to agriculture, based on a classification of support agreed upon by members. Members are expected to provide an annual accounting of their domestic support, including market price support and budgetary expenditures, to be circulated to other members through the WTO Committee on Agriculture (COA).

Although the COA staff facilitate the process through advice and assistance, the individual member governments are responsible for the data and measures presented. Their reports are subject to questioning by other WTO members, but final content is determined by individual members.

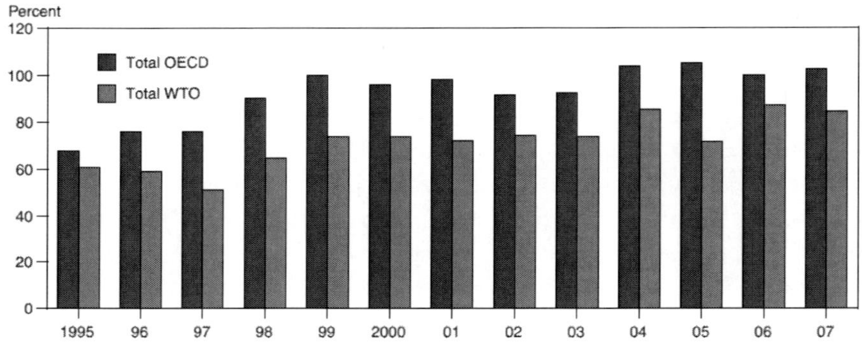


Note: WTO Total Domestic Support includes green box, blue box (1995 only), AMS, and de minimis (support less than 5 percent the value of production) exemptions; OECD Total Support Estimate includes PSE, GSSE, and CSE (see table 2 for definitions).

Source: ERS, using data from OECD PSE/CSE database and ERS U.S. WTO domestic support dataset.

Figure 1. OECD Total Support Estimate and WTO Total Domestic Support for the United States, 1995-2007.

The classification system requires members to categorize their programs according to specified criteria, which identify those programs that may be exempted from reporting as trade-distorting support (table 2). Policies that do not meet the criteria for exemption are by default determined to be trade-distorting, and any support they provide is subject to a maximum support commitment under the URAA. Programs are exempt if they meet the criteria for one of three categories:

- Policies that are at most minimally trade-distorting (“green box”),
- Policies that involve production limits (“blue box”), and
- Policies used by developing country members in the context of broader development programs and that provide development assistance for their low-income and resource-poor populations, as well as support to producers to encourage diversification away from growing illicit narcotic crops (“development box”).

All other policies are subject to maximum support commitments and are reported as part of the Aggregate Measurement of Support (AMS), or the “amber box.” Within the AMS, programs may be reported as product-specific support (associated with production of a specific commodity) or as non-product-specific support (cannot be assigned to specific commodities).

In the final calculation of support subject to a maximum commitment, the *de minimis* rule allows countries to exclude support from the product-specific or nonproduct-specific categories that falls below certain spending limits. For product-specific categories, the *de minimis* limit is 5 percent (10 percent for developing countries) of the value of production of the specific product; for the nonproduct-specific category, the limit is 5 (10) percent of the country's total value of agricultural production. (See WTO, 1994, for the official domestic support notification requirements.)

**Table 2. WTO and OECD main classification categories**

| WTO | OECD |
|---|---|
| • "Amber box" (AMS)<br>— *De minimis* exemptions exclude support less than 5% of the value of production (10% for developing countries)<br>• "Blue box" (production limiting)<br>• "Development box" (development measures—developing countries only)<br>• "Green box" (minimally trade distorting) | • Producer Support Estimate (PSE)<br>• General Services Support Estimate (GSSE)<br>• Consumer Support Estimate (CSE) |

Source: Economic Research Service, USDA, based on WTO *Notifications Handbook* and OECD *PSE Manual.*

## OECD Agricultural Policy Reform Monitoring and Evaluation

The OECD system operates as part of an annual monitoring and evaluation process carried out by the OECD Directorate for Trade and Agriculture. Member countries contribute to the collection of relevant data, but OECD staff are responsible for the reporting process and final calculations, subject to members' review.

OECD's classification system is based on the principle of estimating transfers to producers, both individually and collectively, through policy measures (table 2). The classification system for transfers to individual producers bases categorization on the implementation criteria of each policy, rather than the WTO criteria of whether policies are trade-distorting. OECD members have agreed that the potential trade-distorting impacts of various program designs should not be addressed in the classification scheme (OECD, *Introduction to the PSE and Related Indicators*).

The OECD classification system divides programs into three overarching categories, based on whether the programs are directed:

- to individual producers—the Producer Support Estimate (PSE),
- to producers collectively—the General Services Support Estimate (GSSE), or
- to consumers (first consumers at the farm gate, most often processors rather than final consumers)—the Consumer Support Estimate (CSE).

The total of these three categories produces the Total Support Estimate (TSE). Within each of these larger categories, the classification system further divides transfers based on a hierarchy of implementation criteria.

PSE criteria divide transfers among those based on output; those based on inputs; those based on area, animal numbers, farm receipts, or farm income (A/An/R/I); and those based on non-commodity criteria like environmental benefits. Within the category of A/An/R/I, a further distinction is made based on whether current commodity production is required to receive the transfer, and if so, whether it requires production of a specific single commodity, a group of specified commodities, or any commodity. These categories reflect an effort to organize the classification system to capture members' policy reforms. New transfer programs have moved substantial amounts of support away from traditional output and input subsidies to support based on current production factors and farm revenue, historical production, and non-commodity outputs.

The categorization in the GSSE and CSE is less hierarchical. GSSE categories include expenditures on research and development, agricultural schools, inspection services, infrastructure, marketing and promotion (including foreign food aid), and public stockholding.

CSE categories include transfers associated with policies that raise commodity prices to consumers (the cost to consumers of market price support policies); commodity-specific transfers, such as the share of benefits retained by sugar processors under the sugar loan program; and noncommodity-specific transfers, such as domestic food assistance programs.

OECD goes beyond this initial classification with additional measures of how agricultural support is provided. These calculated measures include:

- Percentage PSE (%PSE), which indicates the share of producer income (gross farm receipts) that comes from transfers from consumers and taxpayers;
- Nominal protection coefficient (NPC), which indicates the share of transfers to producers based on commodity output;
- Percentage single commodity transfer (%SCT), which indicates the share of transfers to producers requiring production of a specific commodity;
- Nominal assistance coefficient (NAC), which indicates the share of transfers to producers from all policy measures; and
- Percentage total support estimate (%TSE), which indicates the share of GDP represented by all measures supporting agriculture.

(See OECD, *PSE Manual,* for further details on these calculated indicators and other aspects of the OECD classification system.)

## CORRESPONDENCE BETWEEN THE SYSTEMS

Despite the apparent similarities between these two systems, there are significant differences across the categorization schemes that limit correspondence between the two systems and their measures.

For example, it might appear that the WTO's green box is roughly equivalent to the OECD's GSSE and CSE, since these categories include many of the same general services and consumer support programs.

But the WTO's green box actually incorporates a number of policies that are classified by OECD in the PSE, including environmental payments, technical assistance, decoupled income support, disaster relief, and some credit programs.

Similarly, some programs that are included in the OECD's PSE and GSSE are not included as part of the WTO domestic support notification at all. Foreign food aid, tariffs, and export subsidies are all reported to the WTO through other notification processes. Moreover, in some cases, the two systems employ different methods of measuring the same type of support. Both the WTO and OECD systems calculate market price support, but their methods produce very different results.

## WTO's AMS vs. OECD's PSE

The comparability across the two systems, or lack thereof, may be shown by consideration of the WTO's Aggregate Measure of Support (AMS) and the OECD's Producer Support Estimate (PSE). Because both measures calculate direct support to producers, it may seem reasonable to assume that the measures are equivalent, or that having access to a country's PSE data would account for its AMS in years for which a country has not yet submitted a WTO domestic support notification.

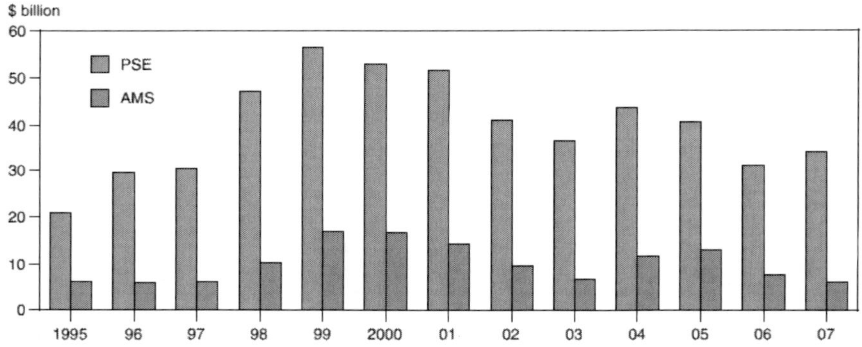


Source: ERS, using data from OECD PSE/CSE database and ERS U.S. WTO domestic support dataset.

Figure 2. OECD's Producer Support Estimate (PSE) and WTO's Aggregate Measure of Support (AMS) for the United States, 1995-2007.

That assumption would be incorrect. Some transfers included in the PSE— such as foreign food aid, tariffs, and export subsidies—are not covered as domestic support under the URAA. Thus, they are not reported as part of the AMS. The difference in total support as measured by these two categories is evident in comparing the PSE and AMS for the United States (figure 2). The difference in methodology for calculating MPS results in reported support differences ranging from $3 billion to $16 billion over 1995-2007. Combined with significantly different methods for classifying direct support to producers, these MPS results contribute to the OECD's PSE ranging from $13 billion to $40 billion higher than the WTO aggregate measurement of support (AMS) over the same period.

The AMS includes all product-specific and non-product-specific support subject to a maximum commitment under the URAA. This includes transfers from higher market prices provided to producers through commodity price support programs. It also includes the value of support to producers from programs tied to production of specific commodities, whether through payments directly to producers or through other benefits like interest subsidies, and the value of payments and other benefits to producers through programs not tied to specific commodities. Support under these latter two categories, however, may be excluded from the maximum commitment ceiling—and thus deducted from the AMS—under the *de minimis* rule (table 3).

**Table 3. OECD's Producer Support Estimate (PSE) and WTO's Aggregate Measure of Support (AMS) subcategories**

| AMS | PSE |
|---|---|
| Product-specific support<br>• Market price support<br>•Direct non-exempt payments<br>•Other non-exempt payments<br>Non-product-specific support<br>De minimis exemptions<br>Any support less than 5% of the value of production (10% for developing countries). For product- specific support, the total must be less than 5% (10%) of the value of production of that commodity; for non-product-specific support, the total must be less than 5% (10%) of the total value of agricultural production. | A. Support based on commodity output<br>1) Market price support<br>2) Payments based on commodity output<br>B. Payments based on input use<br>1) Variable inputs<br>2) Fixed capital formation<br>3) Onfarm services<br>C. Payments based on current area/ animal numbers/ receipts/income (A/An/R/I)<br>D. Payments based on non-current A/An/R/I, production required<br>E. Payments based on non-current A/An/R/I, production not required<br>F. Payments based on noncommodity criteria<br>1) Long-term resource retirement<br>2) Specific non-commodity output<br>3) Other non-commodity criteria |

Source: Economic Research Service, USDA, based on WTO *Notifications Handbook* and OECD *PSE Manual*.

The PSE captures much more product-specific and nonproduct-specific support to producers, as well as other support that under the WTO system is exempt from reporting as part of the AMS (table 3). The PSE classification system incorporates transfers to producers based on output, both through market price support and through per-unit payments. It also includes transfers based on the producers' use of inputs, including variable inputs (fertilizer, fuel, etc.), fixed capital formation (structures, equipment, etc.), and onfarm services (primarily technical assistance). Transfers based on land area, animal numbers, farm receipts, or farm income are also included, and grouped by whether they require current production. Finally, included are transfers related to non-commodity criteria, including resource retirement, non-commodity outputs (like maintenance of certain landscape features), and other non-commodity bases (e.g., flat rate payment per farm).

The OECD's PSE also incorporates a number of policies that may be reported in the WTO green box category of minimally trade-distorting support. These include environmental programs, decoupled income support, and some disaster relief and credit programs. Because the PSE classification principle does not consider the objectives of transfers, programs like environmental payments are classified based on the way they are implemented. Thus, if payments help producers alter their use of variable inputs—limiting fertilizer or pesticide use, for example—the program benefits would appear in the category for transfers based on use of variable inputs. Similarly, if payments are made to producers per animal to support a reduced pasture load, those benefits would appear in the category for transfers based on current animal numbers (A/An/R/I). Decoupled income support, as a transfer to individual producers (in the sense that the land on which it is based remains in agriculture, although it may idled), is included in the PSE and is similarly categorized on the basis of the payment. If support is paid on historical production area or receipts, it will be included in the category for transfers based on non-current area/receipts and not requiring production (A/An/R/I, production not required).

The same contrast between the WTO and OECD systems holds for disaster and credit programs. In the WTO system, members report some types of these programs in the green box, based on criteria for what is considered minimally trade-distorting. In the PSE, benefits are classified according to how the benefit is provided. If the payment is per unit of output, it is classified in the category for payments based on commodity output; if the payment is based on area or animal losses, it is classified in the category for payments based on current area or animal numbers (A/An/R/I).

The implications of these major differences in categorization and inclusion can be seen in a comparison of the AMS and PSE by subcategory for the United States (figure 3). While there is some movement in all support categories over time, the PSE category for support based on commodity output varies more than others and follows a pattern countercyclical to the rise and fall of market prices that drive most of these output programs. Since these are the programs that make up the bulk of the AMS, the two categories track fairly closely, although for the United States the AMS falls below the PSE output support category in most years. This can be explained almost entirely by the difference in measurement methodology for the component both systems call market price support.

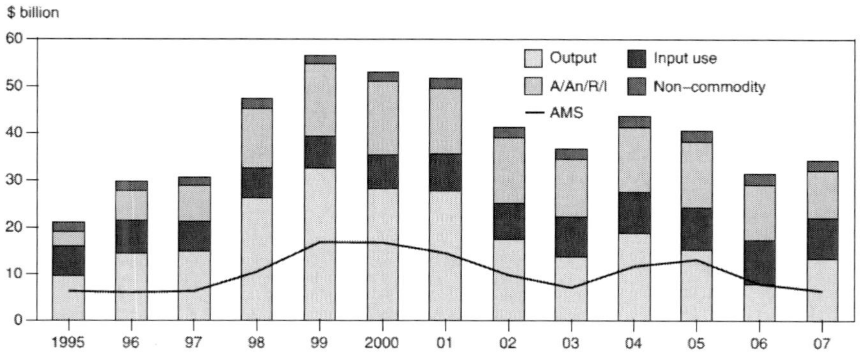


A/An/R/I = payments based on area, animals, receipts, or income.
Source: ERS, using data from OECD PSE/CSE database and ERS U.S. WTO domestic support dataset.

Figure 3. PSE components compared with AMS for the United States, 1995-2007.

## Market Price Support in the AMS and PSE

Market price support (MPS) in both the WTO and OECD classification systems is meant to capture the support provided to producers through measures that raise prices in the domestic market higher than they would be otherwise. These programs are generally operated by limiting imports through tariffs, through government purchases of commodities, and/or the use of export subsidies to support a domestic floor price. Often these measures are used in combination, since maintaining domestic floor prices requires limiting competition from imports to avoid supporting the prices of commodities produced in other parts of the world.

In both the AMS and PSE, the MPS methodology is based on measuring a price gap—the difference between a supported domestic price and the price at the country's borders, or a world reference price. The key difference results from the domestic and world reference prices used by each method and the set of commodities to which the method is applied. The AMS measure of market price support is based on a fixed, historical world reference price—as determined in the URAA in 1994—and applies to commodities for which a country maintains a statutory administered price. The PSE measure is based on current world reference prices and applies to commodities for which there is a measurable gap between the country's current domestic farm price and the world reference price.[2]

The AMS calculation of market price support was developed and agreed to by all members in the URAA. It compares a fixed historical world reference price for the commodity in question to a country's current administered domestic price for that commodity and then applies the difference to the eligible production[3] of that commodity to estimate the support provided to producers:

WTO MPS = (current domestic administered price - 1986-88 average world reference price) * eligible production. (1)

For example, the U.S. dairy MPS for 2007 was calculated as:

WTO MPS = ($218.258/metric ton - $159.825/metric ton) * 85.759 million metric tons = $5.012 billion.

The administered price used in the AMS MPS calculation may not reflect actual returns to producers if it is a floor price that triggers only when market prices fall below that level. Other policies in place such as tariffs and export subsidies may have more impact on producer prices than the administered price. The method does not make explicit reference to any border measures that might be in place in conjunction with the administered price support system—in the WTO, tariffs and export subsidies are notified separately from domestic support (WTO, 1994). Also, if the calculation yields a negative result, the MPS is generally reported as zero.

The URAA includes an alternative calculation when it is not practical to use the price gap methodology, as when no appropriate world reference price can be determined. The methodology for this Equivalent Measurement of Support (EMS) either multiplies the administered price times the quantity of production eligible to receive that price, or reports budgetary outlays used to maintain the producer price at the administered price level (WTO, 2003).

Commodity coverage under the AMS MPS is limited to production for which there is a statutory administered price. For the United States, that means market price support has been calculated for the dairy and sugar price support programs alone since 2002. Before the 2002 Farm Act ended the peanut price support program, peanuts were also included. Because the world reference price is fixed by the URAA, as long as the statutory administered price does not change, MPS for a commodity will vary only with eligible production. For the United States, eligible production of both dairy and sugar[4] has remained fairly constant, which is reflected in the level of AMS MPS (figure 4.)

Market price support (MPS) in the OECD classification system is a more complex calculation, although it is also based on a price gap concept. The method for calculating the PSE MPS for any commodity is to compare the domestic farm-level price with an appropriate world price, usually determined as the price of that commodity at the country's border, adjusted to the farm level. For exporting countries, the appropriate border price is generally the FOB (free on board) unit value; if the country is an importer, the appropriate border price is generally the CIF (costs, insurance, freight) unit value.

The differential, or price gap, is then applied to the country's total production of the commodity. Subtracted from the result are any contributions producers make to the MPS through levies or through "excess feed costs," the share of feed grain MPS that livestock producers have contributed through higher prices for feed (see equation 2).

PSE MPS = [(Domestic farm-level price - border farm-level price) * quantity produced] – producer levies – excess feed costs. (2)

For example, the U.S. dairy MPS for 2007 was calculated as:[5]

PSE MPS = [($421.74/metric ton - $318.40 /metric ton) * 84,233 million tons] – 0 – 0 = $8.7 billion.

The PSE method calculates MPS for 15 common commodities—wheat, maize, other grains, rice, rapeseed, soybeans, sunflowers, refined sugar, milk, beef and veal, sheepmeat, wool, pigmeat, poultry, and eggs—across all OECD member countries. For each country, MPS is estimated for additional commodities based on their share of a country's value of production to ensure that at least 70 percent of the total value of agricultural production is covered. For some countries, MPS may be zero for some of the common and additional commodities. An extrapolation procedure is used to estimate total MPS for use in the PSE:

PSE MPS = (Sum of MPS of individual commodities/sum of value of production (VoP) of individual commodities)

* VoP of all commodities. (3)

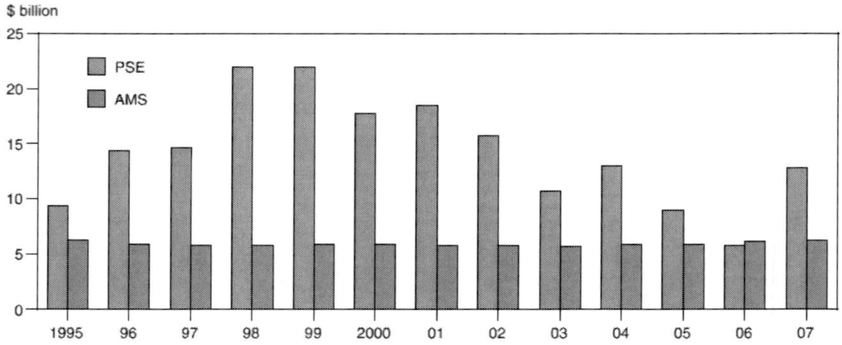


Source: ERS, using data from OECD PSE/CSE database and ERS U.S. WTO domestic support dataset. As noted in the text, the PSE and AMS market price support measures include different commodities. For the United States, the PSE includes wheat, maize, other grains, rice, rapeseed, soybeans, sunflowers, refined sugar, milk, beef and veal, sheepmeat, wool, pigmeat, poultry, eggs, and cotton, while the AMS includes only peanuts, dairy, and sugar before 2002 and only dairy and sugar from 2002 forward.

Figure 4. U.S. market price support as measured by the WTO and OECD, 1995-2007.

As a comparison between the WTO dairy MPS ($5.01 billion) and the OECD dairy MPS ($8.7 billion) highlights, these two methods can yield widely differing results.

Because the OECD method uses the gap between two (domestic and world) current prices to calculate MPS, the amount of market price support varies according to both current prices and current production. For the United States, most of the MPS is accounted for by dairy and sugar, with a small amount for sheepmeat. Since 1995, the U.S. MPS has also included wheat, beef, poultry, eggs, and cotton at much lower support levels than for dairy and sugar.

As a result, the PSE measure, following a pattern countercyclical to the rise and fall of commodity prices, is much more variable than MPS calculated for the AMS. When world prices are high, the gap between the supported domestic price and world price will be small, reducing MPS (as in 2006); when world prices fall, the gap increases and MPS is higher (as in 1999) (figure 4).

In contrast, the MPS calculated for the AMS compares the same fixed world reference price with a stable domestic-administered price, resulting in only slight variation from year to year based on changes in eligible production.

## Comparing Across WTO Exempt Support (Green, Blue, and Development Boxes) and the OECD's GSSE/CSE

While the AMS and PSE encompass some major differences of inclusion and method, virtually all of the budgetary programs included in the AMS appear in the PSE. Other categories of the WTO's domestic support notification are not so easily compared. Blue box programs and development measures are, by definition, support that would otherwise be subject to maximum support commitments under the URAA and included in the AMS, except that they are exempted for specific agreed reasons. Blue box programs are most likely to appear in the OECD's A/An/R/I categories, since production-limiting programs have generally employed area or animal number limits. Development measures would generally appear among the categories for support based on input use (variable input use, fixed capital formation, and onfarm services) for those developing countries for which a PSE is calculated.

The United States has had no blue box programs since 1995, when deficiency payments and acreage reduction programs ended. The United States is not entitled to exempt development measures, but does notify programs in most green box categories.

WTO green box programs are similar to the OECD's GSSE and CSE categories since they include many of the same programs that provide support for "general services" and "consumers" (table 4). However, these categories do not neatly track from the WTO to the OECD classification systems. For example, U.S. program spending reported as WTO green box consistently exceeds the total outlays reported as OECD's GSSE and CSE categories (figure 5). There are several key factors that account for the differences between the OECD and WTO measures. Most importantly, the OECD's CSE captures transfers from consumers to producers that reflect the cost of the higher market prices created through market price support policies. The share of those higher prices, or price transfers, that apply to imported commodities are accounted for as "other transfers from consumers" to differentiate them from transfers to domestic producers. This part of the CSE measure is generally negative, since it is meant to "credit" consumers for policies that create transfers through higher prices. Other components of the CSE include support to consumers from taxpayers, in particular the farm share of domestic food assistance, but also payments to processors. A final component, excess feed costs, accounts for the share of market price support paid for by livestock producers through increased feed prices. In the United States, these latter two components of the CSE have been very low and often zero (figure 6).

**Table 4. WTO's green box and OECD's General Services Support Estimate/ Consumer Support Estimate subcategories**

| WTO | OECD |
|---|---|
| Green box | |
| General services (includes research, extension, inspection, infrastructure, and domestic marketing programs) | GSSE |
| Public stockholding | Research and development |
| Domestic food aid | Agricultural schools Inspection services |
| Decoupled income support Income insurance and safety net Relief from natural disasters | Infrastructure |
| Structural adjustment—producer retirement | Marketing and promotion |
| Structural adjustment—resource retirement | Public stockholding |
| Structural adjustment—investment aids | Miscellaneous (e.g., undifferentiated state-level expenditures) |
| Environmental payments Regional assistance | |
| | CSE |
| | Transfers to producers from consumers |
| | Other transfers from consumers |
| | Transfers to consumers from taxpayers |
| | Excess feed costs |

Source: Economic Research Service, USDA, based on WTO *Notifications Handbook* and OECD *PSE Manual.*

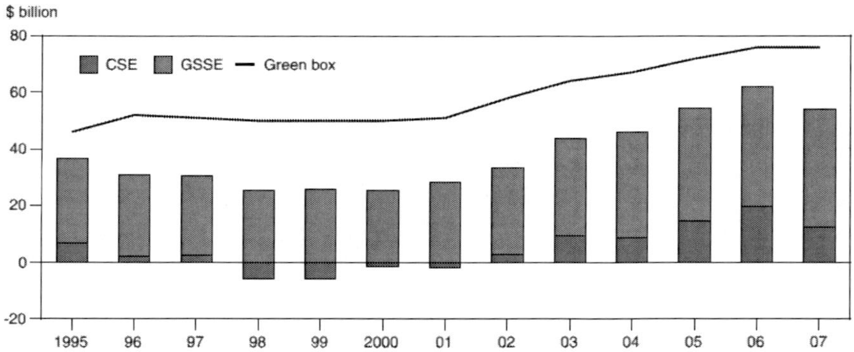


Source: ERS, using data from OECD PSE/CSE database and ERS U.S. WTO domestic support dataset.

Figure 5. General Services Support Estimate/Consumer Support Estimate (OECD) and green box support (WTO) levels for the United States, 1995-2007.

Other differences between the WTO's green box and the OECD's GSSE and CSE totals are accounted for by a number of payments to producers that are exempted from maximum support commitments in the WTO agreement.

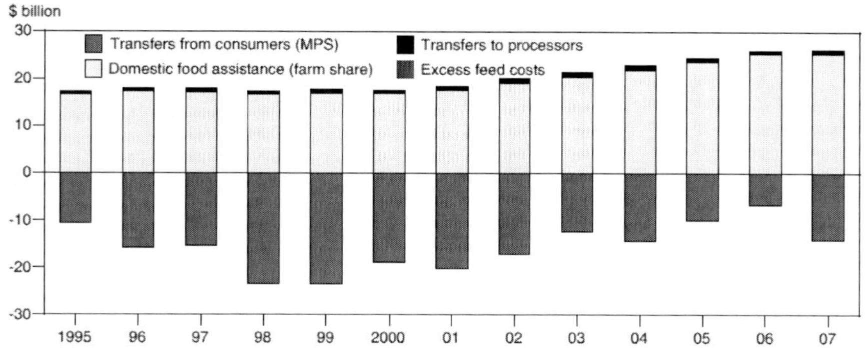


Source: ERS, using data from OECD PSE/CSE database.

Figure 6. Consumer Support Estimate subcategory shares for the United States, 1995-2007.

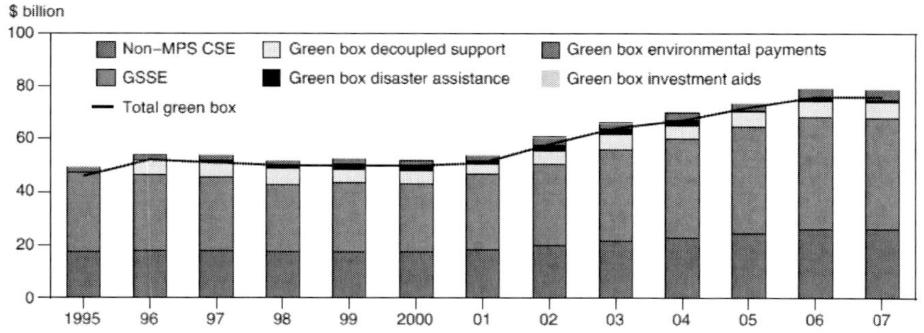


Source: ERS, using data from OECD PSE/CSE database and ERS U.S. WTO domestic support dataset.

Figure 7. General Services Support Estimate, non-Market Price Support Consumer Support Estimate, and selected WTO green box subcategories compared with total green box support level, United States, 1995-2007.

They are reported in the green box because they are considered to have, at most, a minimal impact on trade—decoupled income support, certain types of income insurance and safety-net programs, certain types of disaster relief, certain types of structural adjustment expenditures for producer and resource (usually land) retirement and for investment aids (e.g., credit subsidies), environmental payments, and regional assistance programs. These expenditures appear in the PSE portion of the OECD classification system because they represent transfers to producers. In combination with removing the CSE MPS measure, they account for the difference between the GSSE/CSE and green box totals (figure 7).

## USING THE WTO AND OECD MEASURES OF SUPPORT

Despite many similarities between the WTO and OECD systems for measuring domestic agricultural support, using one system to understand the other or to translate a country's policies from one to the other is not straightforward. Because of the divergent purposes and development of the two systems, key differences in classification and methodologies create significant variations in the measures produced.

In some cases, it may be possible to work carefully from one system to the other to estimate a domestic support measure, either to recreate a missing year of data or to develop some different configurations of domestic support totals that might combine programs and categories of support from both systems. However, translations across systems may not be possible or even appropriate given the potential for misleading results.

Individual programs that make up the WTO domestic support notifications and OECD monitoring and evaluation reports are, in most cases, referred to by the same program names.

For measures that do not vary according to the internal rules of the reporting systems, it is possible to locate and transfer the data on those programs and their expenditures from one system to the other. For some programs, the process simply involves identifying a single program designation in both reporting systems. For others, it requires locating split components of the program and recreating the total for transfer to the other system. Some examples of different types of U.S. agricultural policies and their placement in the two systems illustrate what can be involved (table 5).

Yet even when it is relatively easy to identify and translate programs from one measurement system to the other, a comprehensive understanding of an individual country's policies and programs is needed to identify individual programs within each classification category and apply them appropriately in the other system.

Moreover, the process is only possible for countries whose WTO notifications and OECD monitoring and evaluation data are reported transparently enough to identify these individual programs. Most importantly, some measures, such as MPS and other uniquely calculated indicators, cannot be directly translated from one system to the other and must be estimated for each system according to its own internal requirements to provide consistent and accurate measures.

**Table 5. Examples of OECD and WTO classification of U.S. programs**

| U.S. program | Classification | |
|---|---|---|
| | WTO | OECD |
| Commodity programs<br>Marketing Assistance Loan Program<br>Countercyclical payments | AMS, product-specific support AMS, nonproduct-specific support | PSE, support based on commodity output (output payments)<br>PSE, A/An/R/I (non-current, production not required) |
| Environmental programs<br>Conservation Technical Assistance<br>Environmental Quality Incentives Program<br>Conservation Reserve Program | Green box, general services<br>Green box, environmental payments<br>Green box, environmental payments | PSE, payments based on input use (onfarm services)<br>PSE, payments based on input use (fixed capital formation)<br>PSE, payments based on non-commodity criteria (long-term resource retirement) |
| Other U.S. programs<br>Food Stamp Program<br>Renewable Energy Program | Green box, domestic food aid<br>Green box, general services | CSE, transfers from consumers to taxpayers (noncommodity-specific); GSSE, marketing and promotion (domestic food assistance programs)<br>GSSE, infrastructure |

Source: Economic Research Service, USDA, based on U.S. WTO domestic support notifications and OECD PSE/CSE database.

# REFERENCES


Burfisher, Mary E. (editor). 2001. *Agricultural Policy Reform in the WTO—The Road Ahead.* Economic Research Service, U.S. Department of Agriculture. Agricultural Economics Report No. 802. May.

Economic Research Service (ERS), U.S. Department of Agriculture. *U.S. WTO Domestic Support Reduction Commitments and Notifications Dataset.* Updated Nov. 2009. http://www.ers.usda.gov/Briefing/FarmPolicy/usnotify.htm.

Josling, T.E. 1973. *Agricultural Protection: Domestic Policy and International Trade.* Rome: Food and Agriculture Organization.

Orden, David, David Blandford, and Tim Josling (editors). 2011. *WTO Disciplines on Agricultural Support: Seeking a Fair Basis for Trade.* Cambridge UK: Cambridge University Press.

Organization for Economic Cooperation and Development (OECD). *Producer and Consumer Estimates*, OECD Database, 1986-2008. http://www.oecd.org/document/59/0,3343,en_2649_33797_39551355_1_1_1_37401,00.html.

Organization for Economic Cooperation and Development (OECD). *Introduction to the PSE and Related Indicators*. http://www.oecd.org/document/59/0,3343,en_2649_33797_39551355_1_1_1_37401,00.html.

Organization for Economic Cooperation and Development (OECD). *The PSE Manual*. http://www.oecd.org/document/59/0,3343,en_2649_33797_ 3955 1355_1_1_1_37401,00.html.

Takahashi, Daisuke. 2009. "Quantitative Evaluation of the Japanese Rice Policy Reforms Under the WTO Agreement on Agriculture," *Economics Bulletin* 29(2):712-725.

World Trade Organization (WTO). *Domestic support: Table DS:1 and the relevant Supporting Tables*. http://www.wto.org/english/tratop_e/agric_e/transparency_toolkit_e.htm.

World Trade Organization (WTO). *The Agriculture Committee and official documents*. WTO Agriculture Gateway website: http://www.wto.org/english/tratop_e/agric_e/agric_e.htm.

World Trade Organization (WTO). 1994. *Uruguay Round Agreement on Agriculture*. April. WTO Agriculture Gateway website, WTO Legal Texts page: http://www.wto.org/english/docs_e/legal_e/legal_e.htm#ag.

World Trade Organization (WTO). 1996. *Technical Cooperation Handbook on Notification Requirements*. WT/TC/NOTIF/AG/1. Sept. WTO Agriculture Gateway website, Members Transparency Toolkit page: http://www.wto.org/english/tratop_e/agric_e/transparency

World Trade Organization (WTO). 2003. *Agriculture*. The WTO Agreements Series, vol. 3. (online at WTO Agriculture Gateway website, WTO Agreements series: Agriculture: http://www.wto.org/english/tratop_e/agric_e/ag_notif_e.pdf).

## End Notes

[1]PSEs are produced for the EU and all other OECD member countries, and for some nonmember countries, including Brazil, China, the Russian Federation, South Africa, and Ukraine.

[2] The URAA specifies commodity coverage using the Harmonized System (HS) commodity codes developed by the World Customs Organization. The coverage is broad, including all animal (except fish and seafood), vegetable, and processed food products (HS chapters 1-24), plus edible oils, skins and hides, and raw natural fibers. The OECD focuses on a set of

15 common agricultural commodities for all member countries, plus additional commodities for each country to support actual price gap measurement for at least 70 percent of the total value of production for that country. The full MPS applies the average for covered commodities to the remaining 30 percent of the value of production. Although this difference could lead to more inclusive commodity coverage in the AMS than in the PSE market price support measure, most countries report AMS market price support to the WTO for only a limited number of products.

[3] Member countries decide what constitutes eligible production for their AMS MPS calculations, leading to differences in the way MPS is measured. For example, some countries include all production of the supported commodity, while other countries may include only the quantity purchased by government at the administered price. As a result, if the same commodity is included in both the AMS and PSE MPS, the quantities used may be substantially different.

[4] The United States used total quantity of milk produced as eligible production for the dairy MPS through 2007. For sugar, the total quantity of sugar produced served as eligible production through 2001. After the 2002 Farm Act, eligible production became the smaller of total quantity of sugar produced or the Overall Allotment Quantity (OAQ), which governs the quantity of sugar that can be marketed in a given year.

[5] Eligible production differs between WTO and OECD calculations because of different year definitions. The WTO marketing year for dairy begins October 1 of the notification year (October 1, 2007, for the 2007 domestic support notification). OECD uses the calendar year.

In: Global Agriculture ISBN: 978-1-62618-260-8
Editors: M. R. Robertson, J. I. Schmidt 

*Chapter 2*

# CHINA'S VOLATILE PORK INDUSTRY*

***Fred Gale, Daniel Marti and Dinghuan Hu***

## ABSTRACT

With China's emergence as a new source of potential demand for U.S. pork exports, it is important for U.S. farmers, business leaders, and policymakers to understand the volatile nature of China's pork industry. Prices, hog inventories, and pork output in China fluctuate from year to year in response to various factors that influence the market, and China's imports of pork tend to rise when Chinese hog prices are high. Extensive policy intervention by the Chinese government has contributed to consolidation in the country's pork industry but has not stabilized the market. Imported pork is becoming more competitive in China as Chinese pork production costs rise and animal disease outbreaks, environmental threats, and food safety concerns constrain growth of China's hog industry.

## ABOUT THE AUTHORS

Fred Gale and Daniel Marti are agricultural economists with USDA's Economic Research Service. Dinghuan Hu is a professor with the Institute of

* This is an edited, reformatted and augmented version of United States Department of Agriculture, Publication No. LDP-M-211-01, dated February 2012.

Agricultural Economics and Development, Chinese Academy of Agricultural Sciences.

## INTRODUCTION

China's potential as a major pork importer presents opportunities for hog farmers, business leaders, and investors around the world. Articles and newsletters examining China's effects on the global marketplace reflect buoyant optimism: "The long-run potential for U.S. pork in China is enormous" (Hayes, 2010), and "The potential for further Chinese importation of pork is almost incomprehensible" (a hog industry observer quoted by Dyson (2008)). Announcements of pork sales to China can affect the U.S. market. For example, in October 2009, the *Wall Street Journal* reported "China's pledge to lift a ban on U.S. pork drove prices of lean hogs to a 3-month high on expectations of increased exports to the world's largest pork consumer" (Cui and Waters, 2009).

As China begins to play a larger role in the world pork market, it is important for industry analysts, business leaders, and policymakers to understand the complex factors driving the Chinese hog/pork sector. China's pork industry is constantly buffeted by a range of influences, including disease epidemics, feed prices, policy interventions, seasonal consumption patterns, demand for other meats, and macroeconomic factors. While much attention is focused on the upward trend in commodity prices, pork prices in China tend to rise and fall in multiyear cycles as the industry expands and contracts. The degree of volatility appears to have increased after record-high pork prices in 2007 prompted extensive government intervention and a surge in private investment accelerated structural change in the industry. Following a period of depressed prices in 2010, Chinese pork prices rose to new highs in 2011. China's imports of pork fluctuated in a similar cyclical manner.

This report provides information on volatility in the Chinese pork industry. It reviews recent trends in China-Hong Kong pork imports and fluctuations in Chinese pork prices. It also analyzes the influences of rising feed costs, policy interventions, structural change, and disease epidemics on China's pork industry. As increases in production costs, animal disease epidemics, animal waste disposal challenges, and food safety concerns limit the expansion of China's domestic pork industry, the outlook for pork exports to China is favorable. However, volatility in China's domestic market may result in similar volatility in export sales.

# CHINA'S RISING PORK IMPORTS LINKED TO DOMESTIC MARKET FLUCTUATIONS

China's potential to affect the world pork market derives from the size and volatility of its domestic pork market. China accounts for nearly half of the world's pork production and consumption. Its annual pork output is four to five times that of the United States and more than double that of the European Union. According to official Chinese statistics, China slaughters over 600 million hogs annually—one hog for every 2.2 Chinese people .

Historically, China has been a mostly self-sufficient pork economy. Mainland China traditionally imported modest amounts of pork offal and muscle meats and exported a similar amount of pork and live hogs to Hong Kong (a separate customs territory from mainland China). Some pork shipments from other countries to Hong Kong are re-exported to mainland China through "gray" market channels, but the amount is unknown. While Hong Kong is a short distance from the country producing half of the world's pork, most of the territory's imports come from Europe, the United States, and Brazil. From 2000 to 2006, China and Hong Kong combined to import between 500,000 and 600,000 metric tons of pork and pork products annually.[1] These amounts were a significant share of world pork trade but equated to less than 1 percent of annual pork consumption in China-Hong Kong.

China and Hong Kong pork imports surged in 2007 when a shortfall in Chinese pork production led to record Chinese pork prices. That year, Hong Kong-China pork imports nearly doubled to just over 1 million metric tons (mmt), then rose to over 1 .9 mmt in 2008 (figure 1). According to the U .S. Meat Export Federation, the 2008 total far surpassed the previous pork-import record of 1 mmt set by Japan in 2005.

In 2009, China-Hong Kong pork imports fell to about 1 .5 mmt—still nearly three times the pace of imports earlier in the decade—but rebounded to 1 .8 mmt in 2010. The 2010 import volume was equivalent to 3 .6 percent of China's domestic pork output. The United States supplied about 20 percent of China-Hong Kong pork imports in recent years. For the U .S. pork industry, China-Hong Kong has been one of the leading export markets since 2007. During the peak import period of 2008, U.S. sales to the region accounted for over 18 percent of U.S. pork exports, about double the annual share exported to China-Hong Kong during 2000-2006.

Monthly statistics on mainland China's pork trade reveal a link between imports and domestic pork prices (figure 2).

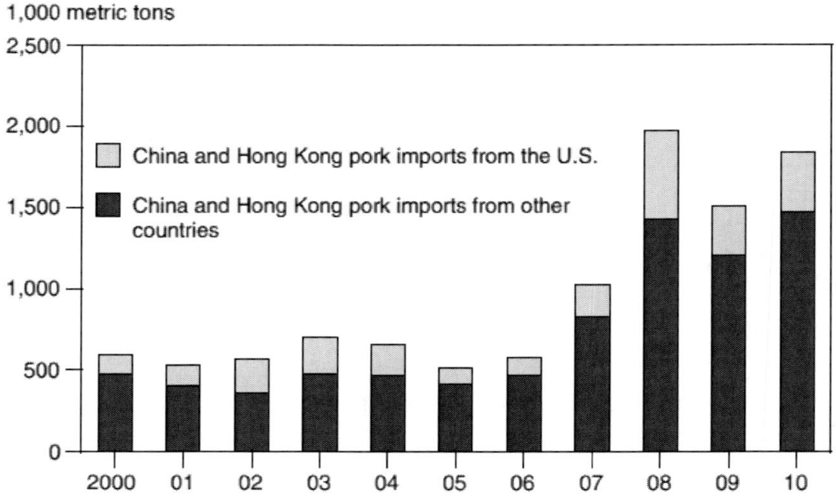


Note: Data include fresh, chilled, and frozen pork; sausage casings; offal; and processed pork products (Harmonized System codes 0203, 020630, 020641, 020649, 0210, 0504, 160241, 160242, 160249).

Source: USDA, Economic Research Service using data from China National Bureau of Statistics and China customs statistics accessed through Global Trade Information Service, Global Trade Atlas.

Figure 1. China and Hong Kong have become a larger market for pork imports.

High domestic prices during 2007-08 reflected short supplies in the Chinese market and prompted a surge in pork imports. The average domestic hog price in China doubled from about $.52 per pound in the first 4 months of 2007 to a peak of $1 .08 per pound in April 2008. Monthly imports by mainland China grew rapidly as prices rose, reaching a peak of 119,000 mt in June 2008, up from 15,000 to 30,000 mt during 2004-06. Imports were boosted by a temporary cut in the country's pork tariff and a state-owned company's contract to purchase U.S. pork to build up reserves ahead of the Olympic Games held in Beijing in August 2008.

After peaking in April 2008, Chinese pork prices fell until early 2009. China's monthly pork imports also fell to under 40,000 mt during the first half of 2009. The decline in prices reached a low of $ .61 per pound in May 2009, about a year after the peak.

A combination of factors helped drive down prices during this period: a buildup in production capacity that increased the domestic supply and a temporary decrease in demand due to concerns among Chinese consumers that the H1N1 influenza virus (swine flu) could be transmitted by eating pork .[2] While no link between pork consumption and H1N1 transmission was scientifically established, Chinese authorities still banned imports of pork from North America to prevent the spread of the disease to China.[3] The ban remained in place for the remainder of 2009 and stopped direct imports of U .S. pork for nearly a year until June 2010.

Chinese pork prices began another run of monthly increases in the second half of 2010 and reached new highs during 2011, 3 years after the sharp increase in prices during 2007-08. With domestic pork prices rising, less expensive foreign pork was more competitive in the Chinese market.

Chinese customs statistics revealed that China's monthly pork imports during late 2010 and 2011 surpassed the record pace set in 2008, rising as high as 150,000 mt during September 2011 (see figure 2). Imports from the United States accounted for most of the import growth during 2011.

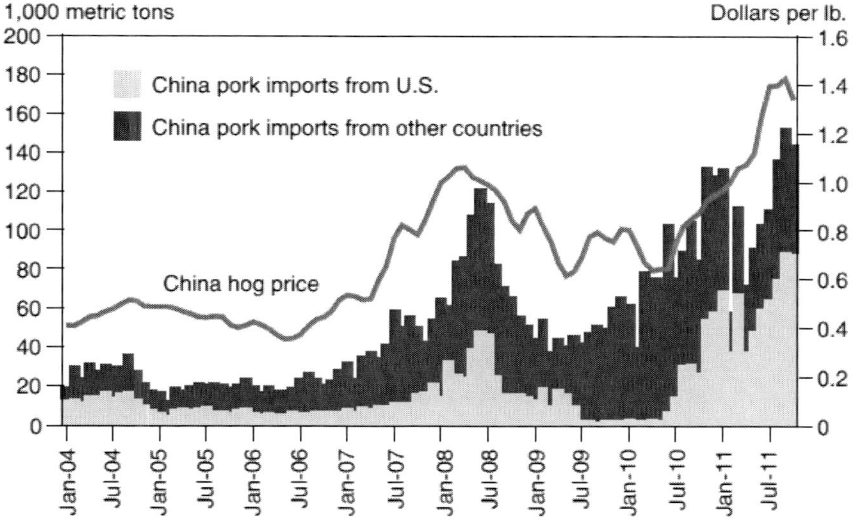


Note: Data include fresh, chilled, and frozen pork; sausage casings; offal; and processed pork products (Harmonized System codes 0203, 020630, 020641, 020649, 0210, 0504, 160241, 160242, 160249). China hog price converted to U.S. dollars at the official exchange rate.

Source: USDA, Economic Research Service using data from China National Bureau of Statistics and China customs statistics accessed through Global Trade Information Service, Global Trade Atlas.

Figure 2. China monthly pork imports vary with domestic hog price.

# RISING PRICES AND COSTS FOR CHINESE PORK

Rising prices in China's domestic pork market were the main contributor to the country's rising pork imports. From 1991 to 2006, Chinese hog prices fluctuated within a relatively narrow range of $ .30 per pound to $ .50 per pound and were usually less than U.S. prices (figure 3). After a sharp increase during 2007-08, however, Chinese hog prices have been significantly higher than U.S. hog prices. While Chinese hog prices fell sharply after peaking in 2008, the average price during 2007-10 ($ .79 per pound) was more than double the average in 1991-2006 ($ .37 per pound). During 2011, prices rose as high as $1 .40 per pound.

The shift in prices is an indication of the general improvement of prospects for U.S. pork sales to China. To be cost competitive in China, however, U.S. pork must be comparable in cost or cheaper than Chinese pork after accounting for freight costs, tariffs (12-20 percent), and value-added taxes (13-17 percent). Overall, we estimate that prices in the Chinese market would have to be approximately 30 to 45 percent higher than U.S. prices for U.S. pork to be cost competitive in China.

U.S. and Chinese consumers have complementary tastes that encourage pork trade between the two countries. U .S. consumers prefer muscle meats, while Chinese consumers prefer offal and variety meats that have low value in the United States (Hayes and Clemens, 1997; Fabiosa et al., 2005). As a reflection of these differing preferences, price competitiveness varies for different cuts of pork. In 2011, the average U .S. prices of livers, hearts, hocks, feet, kidneys, and tails were less than half the prices of corresponding parts in a Beijing wholesale market (figure 4).

Prices of all cuts tend to rise and fall to reflect general changes in pork prices, and in 2011, the rise in Chinese hog prices drove the prices of both muscle meats and variety meats well above corresponding U.S. prices. For example, the price of a lean carcass in Beijing was about $70 higher than the U.S. carcass price of about $91 per 100 lbs. A year earlier, carcass prices had been nearly equal in the two countries. In mid-2011, muscle meat prices were also relatively high in China.

The Beijing price of hams was more than double the U.S. price, the price of spare ribs was nearly 50 percent higher, and the price of bellies was nearly 40 percent higher. Variety meats constitute most of the U.S. pork exported to China, but the widening difference in prices improves the prospects for U.S. muscle meats to be competitive in China.

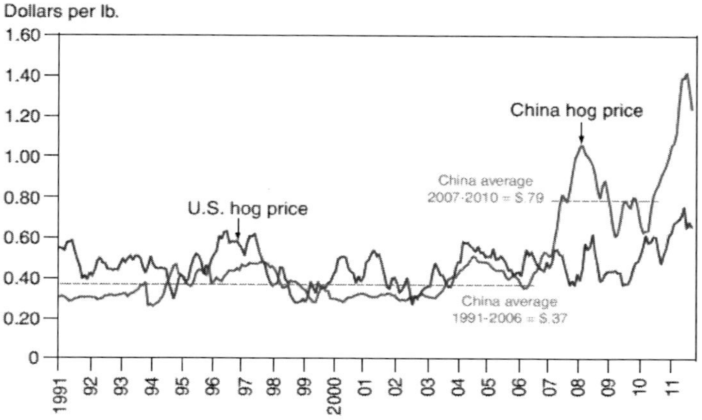


Note: U.S. live hog equivalent, 51-52 percent lean. China price converted to U.S. dollars at official exchange rate. Prices are not adjusted for inflation.

Source: USDA, Economic Research Service using data from USDA, China National Bureau of Statistics, and China Ministry of Agriculture.

Figure 3. The average China hog price has risen above the U.S. price.

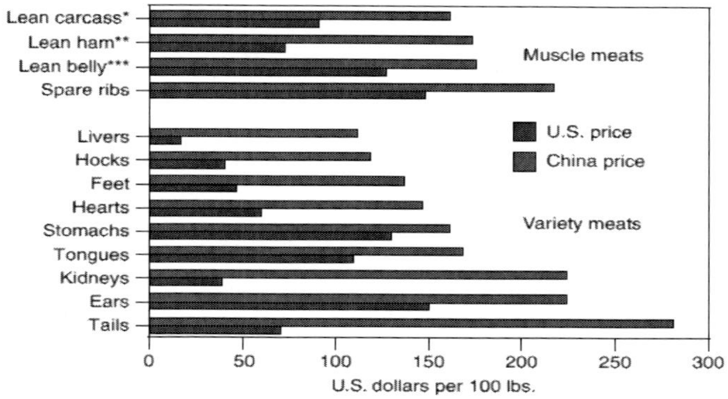


Note: U.S. average price, January-May 2011; China average price from Xinfadi wholesale market, May 2011.

*China lean carcass (bai tiao rou) is compared with U.S. 53-54 percent lean, .65-.80 inches of backfat at last rib.

**China lean ham (hou tun jian) is compared with U.S. composite primal ham.

***China lean belly (wu hua rou) is compared with U.S. composite primal belly.

Source: USDA, Economic Research Service using USDA, Agricultural Marketing Service, LGMN Portal database and Beijing Xinfadi Wholesale Market "price quotations."

Figure 4. Comparison of price of pork parts, U.S. and Beijing averages, 2011.

High pork prices in China likely stem from the rising costs of hog production in China. ERS analysis of data from China's National Development and Reform Commission (NDRC) shows that average hog production costs, converted to U .S. dollars, more than doubled from 2002 to 2009. Chinese hog production costs per pound of live weight rose from about $ .30 in 2002 to $.71 in 2010 (figure 5). Feed is the largest of the hog production expenses in China, accounting for about 60 percent of the total. Feed costs rose from $ .18 per pound in 2002 to $ .25 per pound in 2006. In 2010, the feed cost rose to $.44 per pound, an increase of 77 percent from 2006. The cost of a feeder pig more than doubled during 2002-10 as well.[4] Estimates of hog production costs for 2011 were not available when this report was prepared, but with corn prices rising, it is likely that feed costs rose further in 2011.

Labor and other expenses may have more of an impact on hog production costs than is indicated in figure 5. China's vibrant labor market and an increase in school attendance have absorbed slack household labor that was traditionally used for small-scale "backyard" hog production. According to industry reports, the number of rural households raising hogs has been in decline since 2007. And as these small-scale farms exit the industry, they are increasingly being replaced by larger commercial-scale farms operated by companies or farmers who specialize in raising hogs .[5] Compared with small farms, commercial-scale farms have higher overhead costs for housing, equipment, and manure treatment; they purchase higher value breeds of feeder pigs; and they use paid laborers and technicians instead of relying on unpaid family labor. Commercial-scale farms purchase commercial feeds, while small-scale farms use inexpensive crop stalks, bran and hulls from grains, food scraps, and forages.

The increase in feed costs has pushed Chinese hog production costs above those of the United States. Based on data from China and U.S. production cost surveys for 2009, U.S. hog producers had significantly lower costs per pound of live hog weight ($ .57) than commercial-scale hog producers in China ($.68) and "backyard" producers in China ($ .70) (figure 6). The U.S. cost advantage was mostly due to lower feed expenses and may be a reflection of a more efficient conversion of feed to meat as well as lower feed prices. Even the Chinese cost advantage in feeder pig prices observed in 2002 by Fabiosa et al. (2005) was reversed after the increase in feeder pig prices during 2007. The Chinese surveys may understate costs of commercial-scale producers by excluding very large 10,000-head farms that are becoming more common in China. These farms have overhead costs that may be comparable to the relatively high "other" costs shown for U.S. farms in figure 6.

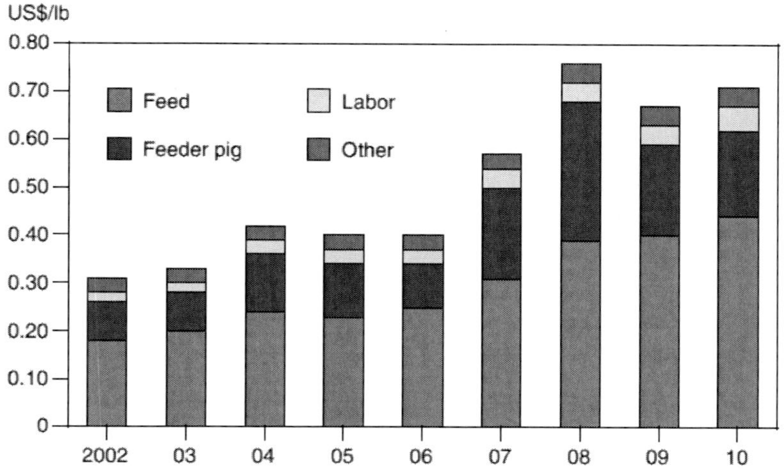


Note: Data in Chinese yuan per head were converted to U.S. dollars per pound of live weight using the official exchange rate. Data are for farms raising 30 or more hogs per year.

Source: USDA, Economic Research Service using data from China National Development and Reform Commission. Data not adjusted for inflation.

Figure 5. China hog production costs are rising.

Grain prices in China have been rising due to the scarcity of cropland and surging demand for grain by feed mills and industrial users. Chinese hog producers and feed mills pay much higher prices for corn than do their U.S. counterparts, and in recent years, rising corn prices have pushed China's feed expenses higher. Rising feed prices tend to push hog and pork prices upward as well. While the Chinese pork industry still uses a wide variety of feeds, Chinese market analysts focus on the price of corn as an indicator of feed costs (see box, "Wide Variety of Feeds Used"). When high corn prices raise the cost of producing pork in China, industry members may take two approaches to alleviate the cost pressure on Chinese consumers: (1) import corn from the United States or other countries with lower corn prices to reduce the cost of producing pork in China, or (2) import pork produced in countries with lower feed costs. While pork production costs are also rising in the United States, U.S. feed costs remain lower than in China because the United States has more abundant land, water, and grain resources. Meeting Chinese demand for pork by producing hogs near sources of feed in the United States and then exporting pork to China is more cost-efficient than exporting large volumes of grain and oilseeds to produce pork in China (Hayes and Clemens, 1997).

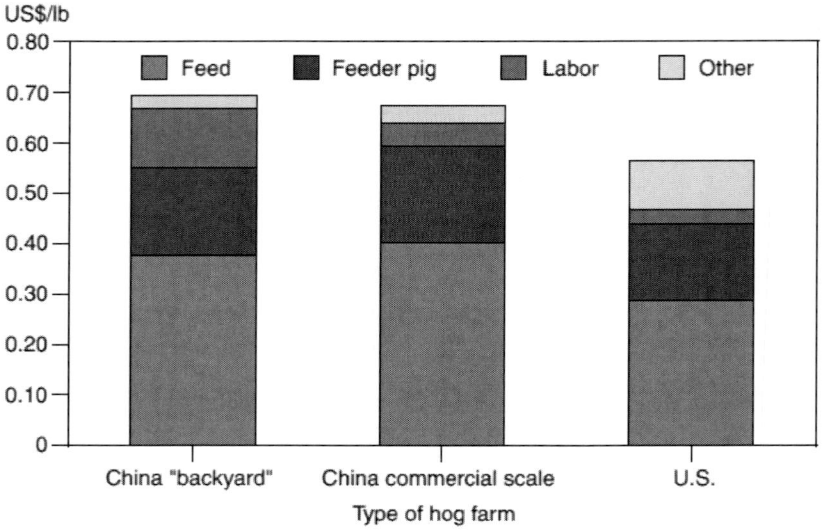


Note: Data in Chinese yuan per head were converted to U.S. dollars per pound of live weight using the official exchange rate. "Backyard" farms raise fewer than 30 hogs per year.

Source: USDA, Economic Research Service using data from China National Development and Reform Commission.

Figure 6. U.S. hog production cost advantage due to lower feed cost, 2009.

The difference between corn costs in China and those in the United States varies from year to year. Corn prices have generally been rising and are higher in China, but surging U.S. corn prices during 2007-08 and 2010-11 narrowed the gap. A comparison of the monthly average cash prices of corn in Guangdong Province and central Illinois during 2005-10 provides evidence of the cost differential (figure 7).[6] The Guangdong corn price increased from about $150 per metric ton to over $300 in late 2010. The price of corn in central Illinois was consistently lower than the price in Guangdong, but the difference between the two fluctuated throughout the 5-year period. In early 2010, the Illinois price was less than half the Guangdong price. The large price difference during mid-2010 stimulated China's first significant corn imports from the United States since the 1990s. (As noted earlier, China's pork imports were also robust during 2010.) U.S. corn sales to China came to a standstill as U.S. corn prices rose and the corn-price difference narrowed later in 2010. However, the Illinois corn price was still 20 percent below the Guangdong price in February 2011.

**Wide Variety of Feeds Used**

A standard Chinese hog feed ration includes about 60 percent corn and 15 percent soymeal. However, the composition of feed can vary widely across farms. Moreover, feed formulations can vary dramatically by region, by farm size, and over time. Chinese farmers have traditionally fed pigs locally available grains, wheat bran, rice hulls, crop residues, vines, potatoes, food scraps, and byproducts from agricultural processors. The composition of feed depends on the types of materials that are available locally at low cost. Farmers often mix these materials with commercial concentrate feeds that contain soymeal, other protein meals, amino acids, vitamins, and trace elements. An increasing number of farmers use commercial formula feeds that include various combinations of the above items already mixed together. Based on interviews with farmers in Sichuan Province, the proportion of corn in hog feed ranges from 30 to 70 percent, while the proportion of commercial concentrate feeds varies from 10 to 20 percent. Commercial-scale farms account for an increasing share of hog production in China and tend to use a higher proportion of corn and commercial feeds than do smaller farms .

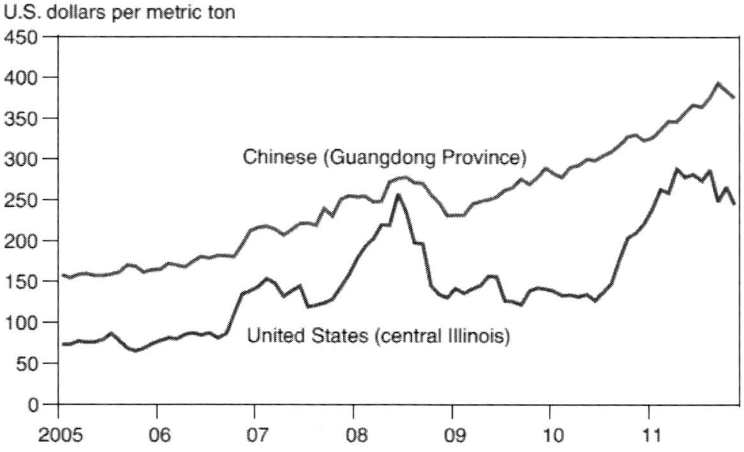


Note: China price converted to U.S. dollars at the official exchange rate.
Source: USDA, Economic Research Service using data from China National Grain and Oils Information Center and USDA.

Figure 7. Corn prices are higher in China.

## CYCLES IN CHINA'S PORK MARKET

The sharp increase in pork prices during 2007 drew attention to the Chinese market's volatility and cycles. Fluctuations in prices and production, however, are not new to the industry. Cyclical patterns in the hog market were recognized in the United States during the 19th century and studied extensively by agricultural economists beginning in the early-20th century (Haas and Ezekiel, 1926). Chinese scholars and analysts have observed similar cyclical fluctuations in China (Liu and Wang 2009; 2010). Zhang (2010a) identified six periods of pork-price increases (1985, 1988, 1994, 1997, 2004, 2007). Several Chinese studies identified a series of four 3- to 4-year cycles in the hog sector during 1996-2009 (Han and Qin, 2007; Liu and Wang, 2009; Nie et al ., 2009). The price increase during 2007 prompted Chinese officials to intervene extensively in the pork market to stabilize prices (discussed later in this report) .

The well-known "cobweb model" was developed in the 1930s to show how cycles in pork prices can result from the biological lag in supply response (Ezekiel, 1938; Coase and Fowler, 1935). In China, it takes 18-20 months to raise a new generation of gilts to breeding age, produce a crop of pigs, and then raise those pigs to slaughter weight (Zhou, 2010). Thus, when an increase in pork prices prompts farmers to expand production, the corresponding increase in pork supply may take more than a year to enter the market. Conversely, when farmers slaughter sows during a downturn in the market, the sector's ability to expand supply is constrained in future months .

Seasonal fluctuations in demand that correspond to major holidays also affect China's pork market. Hog inventories build up in anticipation of peak demand before the Chinese New Year (usually late January or early February) and the mid-Autumn festival and National Day (mid-September to October 1). In the months following these festival/holiday periods, demand, hog slaughter, and pork prices often decline .

Cycles are evident in the ratio of hog price to corn price, an indicator of short-term profitability in the hog sector. A high ratio indicates that the output price (for hogs) is high relative to the price of the chief input (corn), which typically induces farms to build up hog inventories to increase production. Conversely, a low ratio signals financial losses, which prompt farms to decrease hog inventories .

China's hog-corn price ratio has generally fluctuated around 6:1 (figure 8, see box, "The Hog-Corn Price Ratio"). A series of fluctuations in China's pork industry are marked by peaks in the ratio early in 2005 and a much higher peak 3 years later in 2008. Periods of low prices and severe losses for hog producers occurred in mid-2006 and mid-2009, again 3 years apart .

The recent cycles resulted from a chain of events that began in 2004 when concerns over an outbreak of avian influenza drove Chinese consumers to substitute pork for poultry meat. The increased demand for pork increased hog prices and encouraged farmers to expand hog inventories. Hog prices peaked in 2005 but subsequently fell as the supply of pork surged. Prices reached a low in mid-2006, and many producers experienced losses and culled sows. In late 2006 and 2007, animal disease epidemics reduced the supply of pork, and pork prices again began to rise. The previous year's cull of sows and continuing concerns over the effects of animal diseases constrained the industry's ability to expand production in the short run. Demographic changes also affected the pork supply as off-farm employment and rising incomes encouraged many rural families to purchase pork instead of raising hogs in their backyards. The inelastic shortrun supply contributed to a sharp increase in pork prices .

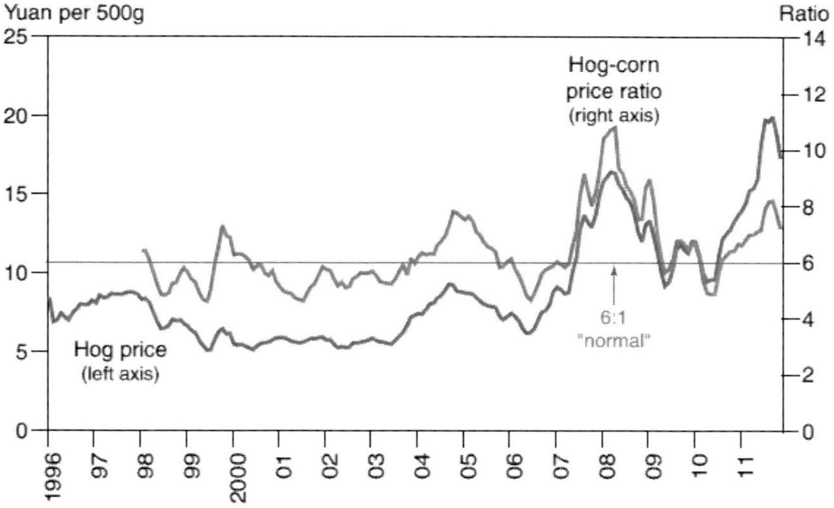


Note: Hog-corn price ratio is a measure of profitability in hog production. Cycles are delineated based on Nie et al., 2009.

Source: USDA, Economic Research Service using data from China National Bureau of Statistics and China Ministry of Agriculture.

Figure 8. China hog prices are cyclical.

Rapidly rising pork prices became a national concern in China during 2007. China's consumer price index for meat was up 40 percent year over year, the highest increase among all categories. The jump in pork prices contributed to the country's high inflation rate that year.[7]

An NDRC study of the factors behind rising pork prices during 2007 focused on the tight supply of feeder pigs. The report found that hog prices received by farmers were up 45 percent year-on-year, but that feeder pig prices were up 60 percent. The feeder pig supply was reduced by the culling of sows and the effects of a disease epidemic that caused sows to abort .

High prices for hogs in 2007 and early 2008 attracted investment in large Chinese hog farms and pork processing. Investors included meat, feed, and real estate companies; overseas investment banks; and a well-known Chinese software entrepreneur (Jiao and Kou, 2010). According to official statistics, China's hog inventory grew 5 percent during 2007, and it expanded at a similar pace in 2008. The hog inventory peaked at nearly 470 million head in late 2009, an increase of 51 million from the 2006 total. This marked China's largest expansion of hog inventories since 1996-99 .

**The Hog-Corn Price Ratio**

The hog-corn price ratio became a valued resource for U.S. commodity market analysts in the 19th century as it reflected the ratio of output price to the price of the main input. At that time, many Midwestern farmers used hogs as an alternative way of marketing corn. When corn prices were low, farmers fed the corn to hogs instead of selling it on the market.

The hog-corn price ratio is now widely used in China as an indicator of hog sector profitability. Market analysts and policymakers in China often view 6:1 as a "normal" level for the ratio. Higher values are considered an indicator of profitability and expansion, while lower values are an indicator of losses and contraction. The breakeven level is often reported to be 5.5:1. The ratio is not as widely used in the U.S. hog industry today because corn accounts for a smaller share of production costs than in years past. The ratio's value as an indicator is also questionable in the Chinese hog industry because corn is only one of many kinds of feeds used by Chinese farmers.

In China, the hog-corn price ratio is computed using prices in yuan per kg. In the United States, the ratio is traditionally computed in dollars per cwt for hogs and dollars per bushel for corn. Because the units used by the two countries differ, the U.S. ratio is typically quoted as a much larger number.

Using data from the 1970s, Van Arsdall and Nelson (1984) calculated a breakeven (cash basis) ratio of 15:1 to 16:1, which would be equivalent to 8 .5:1 to 9 .0:1 calculated using prices in dollars per kg. In a study of historical patterns in the U.S. hog-corn ratio, Holt and Craig (2006) surmised that an increasing trend in the ratio beginning in the 1940s reflected diversification of hog feedstuffs during the post-war period. Recently, U.S. hog market analysts have quoted a higher reference level of 18:1 to 20:1 .

Chinese officials regarded the fluctuations in hog prices and profitability beginning in 2007 as evidence of unusual volatility. This prompted extensive government intervention to stabilize the market. However, a comparison with historical U.S. data for 1909-2010 shows that the recent cycles in the Chinese hog-corn price ratio are not unusual (see figure). Chinese fluctuations are similar to cycles that occurred in the United States during the early 20th century. Coase and Fowler's (1935) description of hog cycles in Great Britain during 1920-33 is also remarkably similar to that of recent cycles in China.

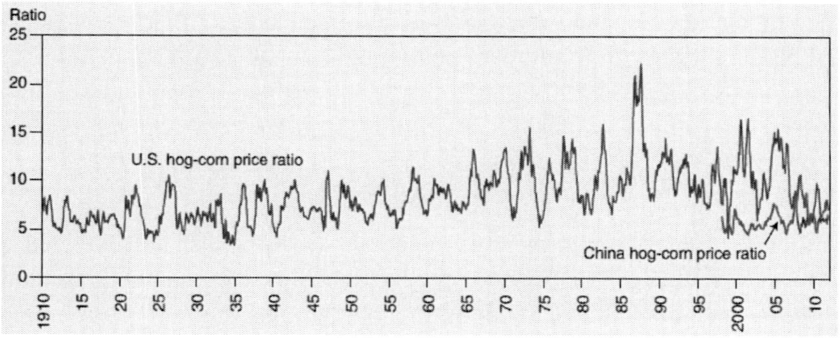


Fluctuations in the China hog-corn price ratio are similar to early 20th-century U.S. fluctuations.

Note: Chart displays monthly ratio of hog to corn price in the United States for 1910-2011 and for China for 1998-2011.

Source: USDA, Economic Research Service analysis of data from USDA, China Ministry of Agriculture and China National Bureau of Statistics.

During 2009 and 2010, two periods of losses and recovery occurred about a year apart, interrupting the pattern of 3-year cycles. The hog-corn price ratio reached a low point in May 2009, about 3 years after the rapid increase in pork prices in 2007. During the second half of 2009, prices rebounded briefly, but the hog-corn price ratio fell again during the first half of 2010 following outbreaks of animal disease epidemics that led to many animal deaths and early slaughter of sick hogs. Furthermore, low prices and economic losses prompted many farmers to cull sows or exit the industry, and China's hog inventory declined during early 2010.

Hog prices rebounded in July that year and continued rising during the second half of 2010 and into 2011. The sector returned to profitability during the second half of 2010, but rising corn prices slowed the increase in the hog-corn price ratio. In early 2011, hog prices neared the historical high reached in 2008, and the hog-corn price ratio was 7:1, slightly above "normal ." Due to high corn prices, however, the 2011 hog-corn price ratio did not rise as high as it did during the 2008 peak period .

By mid-2011, rising pork prices were again a major influence on China's consumer price index, as they had been in 2007. The sow inventory had yet to recover from the losses and culls of the previous year, and the tight supply of piglets slowed the industry's expansion. High feeder pig prices seemed to be consistent with a limited supply of feeder pigs, and Ministry of Agriculture statistics indicated that sow inventories were down 2-3 percent from a year earlier. In one article, industry analysts expressed unease that the situation would be "a repeat of the 2008 roller coaster" (Sun et al ., 2011) .

## CHINESE POLICIES ATTEMPT TO SMOOTH THE CYCLE

Chinese officials use a variety of policy measures that are intended to reduce cyclical fluctuations in pork production. Measures include subsidies, tax breaks, and market interventions in hog farming and pork processing. Government officials often play a coordinating role by recruiting farmers as suppliers, arranging access to land and bank loans, and brokering deals with investors or final customers.

In response to the steep increase in pork prices during 2007, the government announced a package of hog-sector subsidies just 2 months after rising pork prices became a subject of public concern (see Zhu, 2007). A series of additional grants, subsidies, and tax breaks aimed at the pork industry followed later in 2007 and in succeeding years.

A program introduced in 2009 sought to stabilize pork prices by buying and selling pork for government reserves. The government renewed its emphasis on pork policies in mid-2011 when pork prices surged again.

## Pork Industry Subsidies

China's pork policies are part of an ongoing effort to transform the traditional structure of "backyard" farms, small butchers, and pork vendors into a "modern" livestock sector (see box, "Policy Priorities: Modernization and Stability"). Since the 1980s, efforts to modernize the industry have included standardizing hog breeds, feeds, and veterinary medicines; regulating the use of feed additives; enforcing animal health regulations; and shifting toward modern slaughter, processing, and retail markets.

Efforts to modernize China's pork industry were revisited in 2007, a period marked by soaring prices and widespread animal disease epidemics. Officials viewed the prevalence of small-scale backyard farmers as a source of instability because these operations are said to more readily slaughter sows during a market downturn and are more susceptible to animal disease epidemics than larger operations.[8] Officials believe a modernized industry chain will bring stability to the market while also addressing food safety and disease issues.

China's pork policies emphasize improvements in the breeding and farrowing stages of production, upgrading or refurbishing structures and equipment for farms raising slaughter hogs, and attracting investment to the sector.[9] The main policy measures include the following (values converted to U .S. dollars at the official exchange rate in 2010):

- A subsidy payment for each breedable sow set at 50 yuan per head in 2007, raised to 100 yuan ($14 .60) in 2008. The sow subsidy was withdrawn in many areas in 2009 but was restored in 2011.
- Subsidized insurance for sows against losses from disease and natural disasters. The premium (paid to a designated insurance company) is 60 yuan ($8 .78) per head, of which 12 yuan is paid by farmers and 48 yuan is paid by central and local governments. Initially, the program only covered breedable sows, but it was extended to include gilts in 2011. Pilot programs in some areas insure hogs raised for slaughter.
- Free mandatory immunizations against Porcine Reproductive and Respiratory Syndrome (PRRS, called "blue ear disease" in China), foot-and-mouth disease, and classical swine fever. Vaccines are procured from companies through a bidding process with costs split between central and local government. Vaccines are distributed to farms by veterinary officials.

- A "fine breed" subsidy for artificial insemination using semen from boars of approved breeds (such as Duroc, Landrace, and Yorkshire). The subsidy is 10 yuan ($1 .46) for each insemination, for up to 4 attempts per year for each sow.
- Financial awards (grants) of approximately $1 million each to local governments of 362 major pork-supplying counties for financing investments in hog housing, manure handling, immunization, and veterinary work.
- Financial awards to large farms holding at least 500 sows ($146,000) and to standardized farms and village "production zones" where at least 500 hogs are slaughtered annually (from $30,000 to $117,000 according to farm size and province).
- Subsidies of $146,000 were given to each of 300 key breeding farms and provincial hog-breeding centers.
- A waiver of the 25-percent corporate income tax for companies that engage in livestock and poultry production took effect in January 2008 (Petry and Zhang, 2009).

Spending on these programs is only reported on a piecemeal basis by Chinese authorities. The total is difficult to ascertain since there are so many programs, and many are financed jointly by central and local government funds. Central government funds for the "fine breed" subsidy increased from $26 million in 2007 to $95 million in 2010. Pork surplus county awards totaled 3 billion yuan ($450 million) in 2010. Financial awards for large-scale farms totaled $367 million in 2010. Most subsidies are targeted to 362 important pork-producing counties that account for over 40 percent of China's hog production.[10] The award funds are distributed by local authorities to farms, companies, and local officials for refurbishment of hog housing, acquisition of breeding hogs, vaccination programs, manure management, subsidized loans, and support of pork processing, breeding, feed companies and other segments of the industry supply chain.

Various measures are carried out at the local level as experimental pilot programs. These include subsidies for village methane digesters to produce natural gas from hog waste, "ecological" hog-farming projects, subsidized insurance for finishing hogs, funds to pay for disposal of carcasses of diseased hogs, and small cash grants to encourage migrant workers to return to their home villages and set up hog farms.[11] Some of these are extensions of an existing web of local subsidies, model farms, pork reserves, wholesale markets, and supply chain linkages supported by the "vegetable basket responsibility system" in which municipal leaders use various measures to ensure adequate supplies of meat and vegetables for their cities. Pork reserves (both frozen pork and hogs kept in reserve) are held at three levels: central, provincial, and city. In 2011, China's state council ordered large and medium cities in coastal provinces to hold pork reserves to meet 10 days of consumption; other cities were ordered to hold a 7-day reserve.

**Policy Priorities: Modernization and Stability**

In 2006, China's vice minister of agriculture gave a speech at a "Modern Livestock Industry Summit" that typifies the government's approach to livestock industry policy. In his remarks, the vice minister stressed the importance of the hog sector's transformation to a modern industry, citing special instructions issued by the country's top leaders. According to the speech, President Hu Jintao had instructed officials to design policy measures to ensure the stable development of the industry, and Premier Wen Jiabao had called for policies that would stabilize hog prices and prevent large fluctuations in pork production.

Promoting a "modern" livestock sector was emphasized in the government's 5-year plan for 2006-10. Modernization includes the related strategies of "industrialization" (*chan ye hua*), "standardization" (*biao zhun hua*), and increased scale of production (*gui mo hua*). These strategies entail concentrating livestock on larger scale farms, making capital investments in farms and processing facilities, disseminating new technologies and equipment, and integrating farms with processors, breeders, and feed mills.

The vice minister's speech emphasized the need to nurture strong pork processing companies that would establish well-known brands, play a leadership role in the industry, and improve the industry's competitiveness on the international market. He also encouraged farmers to unite in cooperative organizations. The speech called for livestock industry support measures, including subsidies, tax waivers, earmarked bank loans, and other methods. It also said that the government would experiment with methods for attracting private investment to the pork industry.

The vice minister stated that the livestock industry had experienced unprecedented price fluctuations in recent years, and that the hog industry was recovering from a period of very low prices. He emphasized the importance of improving vaccinations and disease resistance. The speech took place about 6 months before the spread of "blue ear" disease and sharp increases in pork prices in 2007. Many of these policies were implemented during 2008-09, but they failed to prevent an even steeper increase in pork prices during 2010-11.

The implementation of hog sector support varies from year to year and place to place. Local government and bank officials make extra efforts to implement policies when orders are issued from national officials. In 2007, the China Bank Regulatory Commission issued a document ordering banks to make loans to expand pork production capacity.

Commercial banks were instructed to lend to companies that raise or slaughter hogs; rural credit cooperatives were instructed to lend to individual hog farmers; and village banks and rural lending companies were instructed to offer production credit. An account by journalists described how local officials in a county of Sichuan Province organized a meeting in 2007 to urge farmers to raise hogs and offer them subsidies while bank officials went door to door to offer loans secured by personal property like ceiling fans and washing machines.

In coastal provinces, hog production is increasingly dominated by large-scale farms operated by companies. In some cities, hog production is banned due to environmental concerns.

In western provinces, slaughter hogs are usually raised by individual small-scale farmers, but officials encourage companies to invest in large breeding and farrowing farms that supply feeder pigs. In 2007 and subsequent years, Ministry of Agriculture officials accelerated implementation of a plan to subsidize construction of "livestock production zones" (*yang zhi xiao qu*) where individual farmers may concentrate their animals in facilities that simulate large-scale farms. In 2011, China's State Council announced a plan to allocate 2 .5 billion yuan ($385 million) annually over 5 years for construction of large-scale hog farms.

In 2007, the combination of policies and high prices attracted investment in new farms, slaughter and processing plants, and imported breeding animals. The buildup of production capacity stimulated by policies helped drive prices down during 2008-09 nearly as fast as they had risen during 2007. Meat sector analysts attributed falling prices to excess supply of pork, with some citing the influx of investment due to government policies as a chief cause (Yi Zhang, 2010, Feng, 2004; Xiao and Wang, 2009).

## "Price Alert" Market Stabilization

In 2009, Chinese policymakers introduced a "hog price alert" market intervention program aimed at reducing the cyclical variation in hog prices. The program's main function is to buy up pork for reserves to increase demand when prices are low and sell pork to augment supply when prices are high. The program intends to stabilize hog inventories by preventing extended periods of low prices that might prompt large culls of sows.[12]

According to documents describing the program, movements in the hog-corn price ratio and several other designated market indicators trigger purchases and sales of pork reserves. The program specifies the "normal" range for the ratio as 6:1 to 9:1, and authorities may take measures to prevent the hog-corn price ratio from falling below 5.5:1.[13] When the ratio falls in various ranges below 6:1, authorities can order designated meat companies to purchase frozen pork to hold in reserve (table 1)[14] When the ratio is 5.5:1 to 6:1, provincial documents specify that cities should maintain a 7-day reserve of pork (based on average consumption of 100 grams/person/day).[15] When the ratio falls below 5 .5:1, the central government can subsidize interest on loans to slaughter and processing companies to encourage them to increase inventories of pork and increase output of processed pork products. Also, the government can authorize an increase in central meat reserves and local reserves in large cities. At "abnormally low" levels below 5:1, the program calls for large increases in reserves, and officials can authorize one-time subsidies per sow for farmers in major pork-producing counties to prevent slaughter of sows.

The document also calls for limits on pork imports "to reduce market supply" and increases in pork exports by raising food safety standards and providing technical, information, and policy support.[16] When the hog-grain price ratio exceeds 9:1, the government can sell frozen pork reserves to bring down prices, and it may issue subsidies to low-income consumers.

Since the program's inception, China's NDRC has published weekly average prices and monthly hog inventory and slaughter on a web site (www.gov.cn/zfjg/szsctk.htm). However, little information about purchases and sales of pork under the program is revealed. News media reports reveal that purchases and sales have taken place, but details on amounts or locations of the transactions are seldom announced.

The first intervention under the stabilization program occurred in May 2009, less than 6 months after its introduction (figure 9). The hog-grain price ratio fell below 6:1, and the government began purchasing pork for reserves in mid-June 2009. During April-June 2010, the government conducted five rounds of pork purchases before prices rebounded in July. Official reports said the hog-grain price ratio fell from 5 .07 in mid-April to 4 .81 by May 26, well into the "red" region specified by the program. It was still in the red region when the fifth round of purchases was carried out at the end of June 2010.

NDRC reported that the hog-grain price ratio reached the 6:1 normal level in late July.

**Table 1. China's pork market intervention guidelines**

| Hog-grain price ratio[1] | Color code | Government action |
|---|---|---|
| Over 9:1 | | Sell frozen pork reserves into the market; issue subsidies to low-income consumers. |
| 6:1 to 9:1 | Green ("normal") | Monitor markets and price fluctuations; issue information. Pork reserves mainly used for emergencies and disasters. |
| 5.5:1 to 6:1 | Blue | Add to central and local pork reserves when ratio is in this range for 4 consecutive weeks. |
| 5:1 to 5.5:1 | Yellow | Subsidize interest on loans to large meat processing companies to encourage them to add to commercial reserves and increase pork processing. |
| Under 5:1 | Red | Increase central reserves and require large and medium cities to increase localreserves of frozen pork when the ratio is in this range for 4 consecutive weeks. The number of live hogs kept in reserve may be increased. If the ratio is still in this range after reserve purchases, a temporary subsidy of 100 yuan per sow may be given to farms in main hog-producing counties when sow inventory is down 5 percent year-on-year. Appropriately limit pork imports to reduce the market supply; "improve" the food safety system to encourage pork exports. |

[1] Some provinces set a higher threshold for the hog-grain price ratio.

Source: USDA, Economic Research Service using National Development and Reform Commission, Regulatory plan for controlling excessive hog price declines, Bulletin No. 1, January 9, 2009, and provincial guidelines for implementation.

Later in 2010, the government ordered sales of pork reserves when pork prices began to rise rapidly. Hog prices rose more than 20 percent during July-November 2010. A round of frozen pork sales was ordered to prevent excessive price increases ahead of the October 1 National Day holiday, and a second round was ordered in November. Some local authorities sold pork reserves in mid-2011 as pork prices reached record levels. These sales occurred when the hog-corn price was far below the 9:1 ratio specified for triggering such sales.

During 2011, as pork prices again became a national concern, there were scattered reports of pork reserve sales by some local authorities but no coordinated national campaign to sell pork reserves. Instead, the State Council issued a new directive ordering local authorities to increase their pork reserves. Large and medium cities and cities in coastal areas were directed to maintain a reserve equivalent to 10 days of pork consumption, and other cities were directed to maintain a 7-day reserve.

The influence of the price alert program on pork prices is hard to discern, especially since the amount, timing, and location of pork reserve purchases and sales are not announced. The program's ability to affect the market is limited because frozen pork purchases constitute a tiny portion of a huge, scattered market.

Zhang and Nie (2010) reported that reserve purchases in 2009 totaled 110,000 metric tons, equivalent to just 0.26 percent of China's annual pork output. The Agricultural Development Bank of China reported that it financed meat-reserve purchases of 250,000 metric tons in 2010, about 0.5 percent of annual pork production.[17] Frozen pork itself constitutes a relatively small part of the market because Chinese consumers have a strong preference for freshly slaughtered meat.[18]

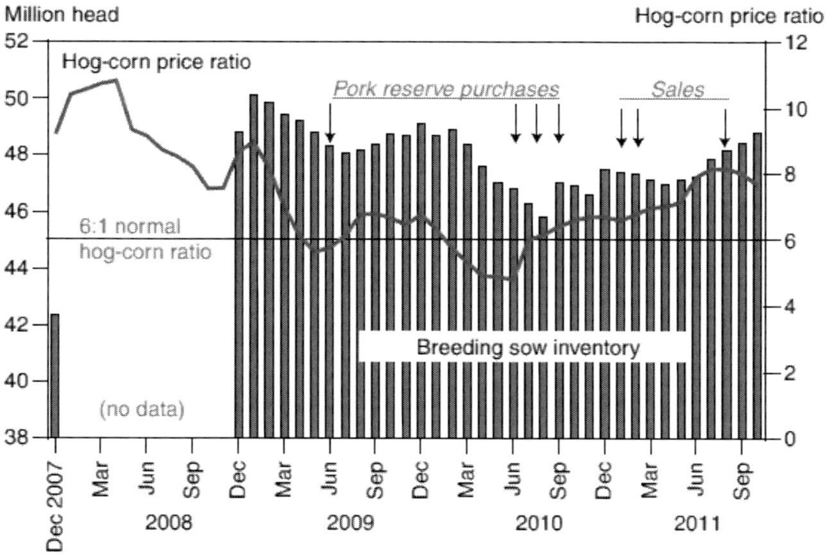


Source: USDA, Economic Research Service using data from China Ministry of Agriculture.

Figure 9. China's sow inventory follows changes in hog-corn price ratio.

Officials credited the first round of purchases in 2009 for bringing about a recovery of pork prices that year, but this seems unlikely since authorities were still recruiting companies to hold reserves and many provinces had not even released their implementation regulations at that time. Some market reports again credited reserve purchases for the rebound in hog prices in July 2010, but the rebound was more likely due to a 5.7-percent decrease in hog slaughter that month.

The weakness of the hog price alert program was revealed by the sharp increase in pork prices during 2010-11. During this period, rising food prices were a major policy concern in China and officials ordered sales of pork in the fall of 2010, yet hog prices rose 40-to-50 percent and by June 2011 pork prices reached the record level set in early 2008.

Market reports ascribed the surge in prices to short supplies of pork, which resulted from widespread animal disease, culls of sows, and exits from the industry during the period of low prices and losses in 2010. The supply of feeder pigs in 2011 was limited by the cull of sows a year earlier, while rapid income growth created robust demand (Woolsey and Zhang, 2011). The wide swing in prices and hog inventories during 2010-11 was remarkably similar to that of 2006-07—exactly the type of phenomenon that the price alert policy was intended to prevent.

## SWINE EPIDEMICS AFFECT THE PORK MARKET

Outbreaks of "blue ear" disease (PRRS), foot-and-mouth disease (FMD), classical swine fever, pneumonia, streptococcus suis, circovirus, parasites, and erysipelas (a bacterial infection of the skin) are common in China's hog industry. News reports indicate that large losses attributed to disease periodically restrict the supply of pork, contributing to price surges. Chinese news media periodically report illegal sale of pork from diseased hogs, pork that is discolored or bearing lesions, and discovery of large numbers of hog carcasses in rivers and canals.

The impact of disease on the pork market is impossible to assess with any precision. When disease affects sows and young pigs, impacts on the market may occur several months after an outbreak. The frequency and incidence of disease is hard to measure accurately, and outbreaks are often regional. While the Ministry of Agriculture reports on disease outbreaks, these figures likely understate the actual incidence of disease because farmers and merchants have little incentive to report diseased animals to authorities (Woolsey et al., 2010).

Rising pork prices in 2007 and 2011 were attributed in part to disease outbreaks. During 2006-07, the supply of feeder pigs was reduced in part because blue ear disease caused sows to abort. This restricted the supply of finished hogs and led to high prices during 2007-08. During 2010, another round of disease outbreaks induced many farmers to slaughter hogs early, and many diseased carcasses were illegally sold to slaughterhouses. The increased supply of slaughtered hogs and the prevalence of tainted pork led to a period of depressed prices during 2010 (see Woolsey et al., 2010). The disease-related decline in hog inventories during 2010 led to tight supplies of feeder pigs and another surge in pork prices during 2011 (Woolsey and Zhang, 2011; Sun et al., 2011).

Chinese agricultural officials have taken steps to reduce the effects of animal disease epidemics. Immunizations for PRRS, classical swine fever, and FMD disease are compulsory and subsidized. Farmers are compensated for culling animals to prevent epidemics, and local officials must bear the cost of sanitary disposal of dead animals. Agricultural officials have made it a priority to improve veterinary services, biosafety, and monitoring and control of farms, slaughterhouses, transportation, and markets. Chinese pigs are required to have ear tags recording 7 to 8 mandatory immunizations. Slaughterhouses are required to ensure that hogs have ear tags and vaccination certificates. Inspectors are required to conduct ante- and post-mortem inspections of slaughtered hogs, and tissue samples are to be examined for parasites at slaughterhouses.

However, government programs and regulations are not uniformly implemented, and problems with animal health persist. Vaccines may not be effective against multiple strains of PRRS that exist in China's hog herds. Wang (2009) reported that ear tags (that record immunizations) and animal quarantine certificates can be easily falsified or purchased; farmers often miss vaccinations; syringes are often used on multiple animals; and veterinary personnel needed to carry out vaccinations are in short supply. In a small survey of farmers, Liu et al. (2007) found most farmers whose sick pigs did not respond to treatment sold the animals before they died or improperly discarded the carcasses of dead animals (only one farmer said he buried dead pigs as required by regulations). Liu et al. found that most farmers did not use ear tags or only attached them at the time the pigs were sold; pigs without ear tags were sold to unlicensed butchers or neighboring farmers.

The high density of animal populations, rudimentary facilities, lack of technical knowledge among farm personnel, and lack of resistance among "foreign" breeds of pigs may contribute to vulnerability to disease. One article in a Chinese veterinary publication warned farmers that changes in feed, poor nutrition, neglect of immunizations, and other animal stress-inducing factors that are more common on small-scale farms may heighten risk of classical swine fever outbreaks (Ren, 2010). According to Feng (2010), periods of extreme weather or flooding can trigger an epidemic among farms characterized by stressed animals and neglect of immunizations .[19] As evidence of farmers' cost-cutting, Feng noted that many feed and veterinary drug companies reported sales declines of as much as 30 percent during the period of losses in 2010. Feng also suggested that periods of losses by farmers may leave local governments short of cash to pay veterinary technicians and dispose of carcasses. Feng (2010) hypothesized that disease outbreaks may actually be linked to the hog cycle itself. Writing in an industry newsletter, he observed that extended periods of losses for small- and medium-scale hog farmers may leave them short of cash.

Farmers may then cut costs by switching to lower quality, less-nutritious feeds, and neglect vaccinations. Less nutritious diets leave animals in a weakened state and vulnerable to disease. Jiao and Kou (2010) reported that farmers in Sichuan cut costs during a period of depressed prices by increasing the proportion of rapeseed stalks in hog rations. Sun et al. (2011) observed that the substitution of cheaper feed substitutes contributed to chronic disease problems in the industry.

Chinese officials introduced subsidized insurance for producers of sows to reduce the risk of financial losses attributed to animal deaths, but some market reports indicate that this program encountered moral hazard problems. Liu (2010) reported heavy losses incurred by companies offering sow insurance in Henan Province, citing large numbers of claims, high costs, and fraud .[20] In July 2010, a Yangzhou newspaper reported that the insurance company's payouts for sow death losses had doubled from the previous year and reached 25 percent in one county (Yang, 2010). According to an insurance worker quoted in the article, some farmers neglected to treat sows for diseases during periods of low hog prices because the insurance indemnity exceeded the animal's salvage value if it were culled and sold.

## ENVIRONMENTAL AND FOOD SAFETY PRESSURES

Producing large quantities of pork in China entails "external" costs—the environmental impacts of hog waste and risks of food safety incidents— that are not factored into the market price of pork. According to Wang et al. (2006), one Chinese hog produces 5 .3 kg of waste daily, which contains large amounts of nutrients not absorbed by the animal as well as heavy metals and pharmaceutical residues. During the 1950s and 1960s, Chinese officials encouraged individual households to raise hogs as a means of producing organic fertilizer to spread on fields to raise grain yields. In later decades, chemical fertilizer became available and the rising demand for meat prompted a large increase in the number of hogs. Consequently, the production of manure exceeded the capacity of the surrounding farmland to absorb its nutrients. Small farms rarely treat manure, but large farms are usually required to invest in treatment facilities. Gao et al. (2006) estimated that 80 percent of commercial-scale farms lack equipment and facilities to properly dispose of waste, which causes "serious pollution of water, soil, and air and threatens the health of animals and humans ." Waste often washes into streams and rivers, fouling drinking water and contributing to eutrophication (nutrient enrichment) of major bodies of water .

Several Chinese studies estimated that pollution from livestock farms totaled roughly 3 billion metric tons annually, about three times the pollution emitted from industrial sources (Gao et al ., 2006; Liu, 2009; and Wang et al ., 2006). Gao et al. estimated China's hog waste at 1 .29 billion metric tons annually, 47 percent of the total livestock and poultry waste generated. A census of pollution sources released in 2009 found that livestock waste was a chief cause of water pollution in China (China Ministry of Environmental Protection, 2010).

While Chinese officials are taking steps to address these problems, the dense population of hogs strains the capacity of the land to supply feed for hogs and absorb their waste and also makes it difficult to control and prevent animal diseases (FAO, 2006). China has many regions with high hog population densities. ERS calculations using provincial- and state-level data on hog inventories and cropland for China and the United States show that China had 94 hogs for every 100 acres of cropland nationwide at year end in 2008, more than 4 times the U.S. ratio of 20 hogs per 100 acres (table 2). In many of China's leading hog-producing provinces, the density exceeded 100 hogs per 100 acres, and the density exceeded 200 in Sichuan, Hunan, and Guangdong Provinces. North Carolina was the only U.S. State with over 200 hogs per 100 acres (similar to the density in Sichuan), and Iowa had 82 (slightly below the China average and similar to the density in Hebei and Jiangsu Provinces). Other U .S. States had densities of fewer than 40 hogs per 100 acres, far less than in China. With an already-high density of hogs, the environmental impact of hog production and tight supplies of feed may constrain growth of China's hog industry.

**Table 2. Comparison of hog population density in China and United States, 2008**

| | Hog population | Hog-cropland density | | Hog population | Hog-cropland density |
|---|---|---|---|---|---|
| | *Million head* | *Hogs per 100 acres* | | *Million head* | *Hogs per 100 acres* |
| China | 462.9 | 94 | | 66.7 | 20 |
| Leading provinces: | | | Leading States: | | |
| Sichuan | 53.3 | 221 | Iowa | 19.8 | 82 |
| Henan | 44.6 | 139 | North Carolina | 9.6 | 225 |
| Hunan | 39.2 | 255 | Minnesota | 7.4 | 37 |
| Shandong | 27.3 | 90 | Illinois | 4.4 | 19 |
| Yunnan | 26.7 | 109 | Indiana | 3.5 | 29 |
| Hubei | 24.6 | 130 | Nebraska | 3.4 | 17 |
| Guangdong | 23.8 | 206 | Missouri | 3.1 | 23 |
| Guangxi | 23.1 | 135 | Oklahoma | 2.4 | 26 |
| Hebei | 20.2 | 79 | Ohio | 1.9 | 19 |
| Jiangsu | 17.2 | 89 | Kansas | 1.7 | 7 |

Note: Hog population is year-end inventory. Density is the ratio of hog inventory to total cropland. Provinces and States are ranked by hog population; only the top 10 are shown.

Source: USDA, Economic Research Service using data from China National Bureau of Statistics and USDA, National Agricultural Statistics Service.

Chinese officials are promoting "ecological" modes of hog production that use hog waste to feed fish or fertilize crops and use bacteria to break down hog waste. China's 2011-2015 5-year plan will emphasize the importance of controlling livestock waste (Zhang, 2010b). However, a number of cities and provinces in China have introduced regulations that ban hog farms and slaughterhouses from operating near residential areas and waterways. Draft regulations prepared by Shandong Province in 2010 banned new livestock farms in urban areas, near sources of drinking water, in scenic areas, and in places where toxic substances exceed prescribed limits.

Food safety is also a major concern for Chinese consumers of pork. The news media in China has frequently reported on the hog industry's use of clenbuterol and other illegal feed additives, the slaughter of sick hogs, the pumping of potentially contaminated water into hogs prior to slaughter, and the contamination of feed with heavy metals. Chinese consumers are also becoming more wary of pork products that contain dyes, preservatives, and other food additives. In March 2011, a widely publicized report that a subsidiary of China's largest processed pork manufacturer purchased hogs raised with illegal feed additives had little direct impact on the pork market (Woolsey and Zhang, 2011). However, industry reports claim that the incident helped drive the trend toward consolidation of hog farms discussed earlier in this report. For example, the company implicated in the incident pledged to open a 10,000-head company-operated farm to supply each slaughterhouse it builds to gain more control over the production process.

Food safety concerns are also contributing to changes in purchase patterns that may make consumers more receptive to imported pork. Traditionally, Chinese consumers preferred to purchase freshly slaughtered pork from small wet market vendors, but food safety concerns have encouraged them to shift purchases to supermarkets where pork is believed to be more sanitary and free of illegal feed additives. Government plans to consolidate slaughterhouses by 2015 entail an increase in interregional trade in chilled or frozen pork. The diminishing role of localized wet markets and the development of modern market channels with cold chain facilities may create more opportunities for imported pork to reach Chinese consumers. Many Chinese consumers responded to the dairy industry's melamine adulteration crisis by purchasing imported milk products, and demand for imported pork could similarly be boosted by domestic food safety concerns.

## LOOKING FORWARD

China's status as a major pork importer will likely continue to grow. China's tradition of self-sufficiency in pork will be hard to maintain as feed costs rise and as land for expanding farms and processing facilities becomes scarce and expensive. The environmental and food safety impacts of producing large numbers of hogs in China will become more apparent. Interregional shipments of pork within China are limited by lack of reliable transportation and temperature-controlled storage.

Stricter regulatory enforcement in the United States, greater investments in animal housing and manure handling, wider dissemination of technical expertise, and closer coordination between producers and processing companies help U.S. farmers produce pork with less of an impact on the environment, fewer food safety incidents, and fewer disease outbreaks than in China. Demand from China raises the value of variety meats and offal that are not widely used as food in the U.S. market. Chinese restaurant chains, hotels, and other buyers who demand pork with high and consistent quality are important potential customers for imported pork (Fabiosa et al ., 2005). With diversifying consumer tastes and growing segmentation in the market, imported pork can coexist in the Chinese market with domestic grain-fed pork and meat from local pig breeds.[21]

Strong resistance to pork imports in China can disrupt trade and affect exporters. China lowered tariffs on pork after its accession to the World Trade Organization, but pork imports still face resistance similar to that described by Hayes and Clemens (1997). Evidence of this can be seen among the policy responses listed in the Chinese "hog price alert" program, which include unspecified "limits" on imported pork to reduce the market supply and "encouragement" of pork exports. When Chinese pork prices were soaring in 2007, officials made announcements to assure the public that China would not import large amounts of pork (Xinhua, 2007).[22] In June 2010 (after China lifted its H1N1-related ban on U .S. pork), an article entitled "Be on Guard! American Pork's 'Soybean Appetite'" warned that imports could "eat up" China's pork industry if the industry was not protected (Li, 2009). An analyst quoted in the article cautioned readers to "Be careful of the trap set by the Americans," warning that if U .S. pork imports are not limited, the pork industry "...is likely to repeat the mistakes of the soybean industry with disastrous consequences."[23] Similarly, Liu (2010) reported that Chinese officials were wary of foreign investment in the pork industry because officials feared losing "guidance power" over the industry.

## References

China Animal Husbandry Information Center. "Jiedu Di Wu Ci Shouchu 'Zai Du Chu Shou' Ken Hou Yuanyi (Interpreting Reasons Behind the Fifth Purchase of Reserves) ." July 2, 2010. http://livestock .cn/

China Ministry of Agriculture. *Livestock Industry Yearbook.* Beijing: China Agriculture Press, various years .

China Ministry of Environmental Protection, National Bureau of Statistics, and Ministry of Agriculture. *Di Yi Ci Quan Guo Wuran Yuan Pucha Gongbao* (Communique of the First National Census of Pollution Sources). Posted by Xinhua News Service, February 9, 2010. http://news. xinhuanet .com/politics .htm

China Ministry of Finance. *Ministry of Finance Notice on Rising Pork Prices and Promotion of Healthy Development of the Hog Industry* (in Chinese). No. (2007)221, June 18, 2007 .

China National Development and Reform Commission (NDRC)/Ministry of Agriculture. *Declaration of Hog Standardization, Large Scale Farms (Zones), Hog and Poultry Breeding Construction Project Plan* (in Chinese). NDRC Rural Economy No .(2008)524, March 3, 2008.

China National Development and Reform Commission (NDRC), Office of Price. "2007 Nian Shang Ban Nian Shengzhu Chengben, Shouyi, Jiage Jun Chuang Lishi Xin Gao (Hog Production Cost, Profit and Price for the First Half of 2007—Price Reaches a New Historical High) ." September 5, 2007. http://jgs .ndrc .gov .cn/jgqk/t20070905_157541 .htm

Coase, R .H ., and R .F. Fowler. "Bacon Production and the Pig-cycle in Great Britain," *Economica*, May 1935, pp. 142-167.

Cui, Carolyn, and Theopolis Waters. "Hog Prices Hit 3-Month High as China Vows to End Pork Ban," *Wall Street Journal*, October 30, 2009. http:// online .wsj .com/article/SB125682230437515867 .html

Dyson, Tom. "How We'll Make a Fortune Exporting Pork to China," *Daily Wealth*, July 30, 2008. www .dailywealth .com

Ezekiel, M. "The Cobweb Theorem," *Quarterly Journal of Economics*, Vol. 53, 1938, pp. 255-80 .

FAO (Food and Agriculture Organization of the United Nations). *Livestock's Long Shadow: Environmental Issues and Actions.* Rome: FAO, 2006 .

Fabiosa, J ., Dinghuan Hu, and Cheng Fang. *A Case Study of China's Commercial Pork Value Chain.* MATRIC Research Paper 05-MRP 11, Iowa State University, August 2005. http://www .card .iastate .edu/publications/synopsis .aspx?id=788

Feng, Yuhui. "The Hog Market Welcomes Its Longest Profit Cycle in 10 Years" (in Chinese). China Feed Industry Association, unpublished market analysis, 2004. http://www .xqyzz .cn/file/sc05 .htm

_____. "An investigation of hog and sow inventory changes in several regions" (in Chinese), *China Journal of Animal Science*, June 2006 .

_____. *Soozhu China Pork Market Analysis and Warning Report* (in Chinese). Zhong Ke Yi Heng Modern Agriculture Information Institute (Beijing) weekly newsletter, June 2010 .

Gao, Ding, Tongbin Chen, Bin Liu, Yuanming Chen, Guodi Zheng, and Yanxia Li. "Releases of Pollutants From Poultry Manure in China and Recommended Strategies for the Pollution Prevention" (in Chinese), *Geographical Research*, 2006, Vol. 25, No. 2, pp. 311-319 .

Haas, G .C ., and M. Ezekiel. *Factors Affecting the Price of Hogs.* U .S. Department of Agriculture Bulletin 1440, 1926 .

Han, Jie. “Finance Ministry Allocated 6 .5 Billion Yuan in Funds for Sow Insurance and Subsidy” (in Chinese). Xinhua News Agency, June 26, 2007.

Han, Jun, and Zhongchun Qin. “Analysis of Our Country’s New Round of Hog Cycle Fluctuations” (in Chinese). Special manuscript, State Council Development Research Center, August 2007, and *Today’s Animal Husbandry and Veterinary Review*, September 2007.

Hayes, D .J. “The Potential Market for U .S. Pork Exports in China,” Iowa State University Working Paper, January 13, 2010 .

Hayes, D .J ., and Roxanne Clemens, “The Chinese Market for U .S. Pork Exports,” Iowa State University CARD Briefing Paper 97-BP 14, March 1997.

Holt, M .T ., and L .A. Craig. “Nonlinear Dynamics and Structural Change in the U .S. Hog-Corn Cycle: A Time-Varying STAR Approach,” *American Journal of Agricultural Economics*, Vol. 88, No. 1 (February 2006), pp. 215-233.

Jiao, Jian, and Bo Kou. “Wen Luan ‘Zhu Zhouqi’ (Disorderly ‘Hog Cycle’)” (in Chinese), *Caijing Magazine*, No. 13, 2010.

Li, Jinling. “*Jingti! Meiguo Zhuroude ‘Dadou Weikou*’” (Be on Guard! American Pork’s ‘Soybean Appetite .’) *Zhongguo Chanjing Xinwen* (China Industry and Economic News), May 26, 2009. http://finance .sina .com .cn/ roll/20100526/18538007356 .shtml

Liu, Fang. “*Zhengce Zhu PK Shichang Zhu: Wenluan Zhouqi* (Policy Hogs Replace Market Hogs: Disorderly Cycle),” Dahe Henan Provincial News Network, July 26, 2010. http://newpaper .dahe .cn/dhcf/page/402/2010-08- 03/16/716412807919372 81 .pdf

Liu, Chunfang, and Jimin Wang. “Analysis of Recent Fluctuation in Our Country’s Hog Prices and Future Trends” (in Chinese). Institute of Agricultural Economics, Chinese Academy of Agricultural Sciences, Research Briefing No. 218, May 20, 2009, and *Agricultural Outlook*, May 2009.

_____. “Status Quo and Prospect of Pig Industry Development in China” (in Chinese), *Agricultural Outlook*, Vol. 6, No. 3, March 2010, pp. 28-31.

Liu, Heguang, and Dongsheng Sun. “Development of China’s Pig Meat Consumption and Its Prospect” (in Chinese), *Agricultural Outlook*, Vol. 6, No. 1, January 2010, pp. 35-38.

Liu, Yuman, Xiaoqing Yi, Yintong Du, and Lei Wang. “Issues of Food Quality and Safety in the Supply of Pork Chain,” *Chinese Journal of Animal Science* 43(2), 2007.

Liu, Zhaozheng, "Study on Current Rural Environmental Problems," *Issues in Agricultural Economics*, 2009, Vol. 30, No. 3, pp. 70-73.

Nie, Fengying, Ling Dong, and Jieying Bi. "Fluctuation and Cycle of Pork Price in China ." Selected paper presented at International Agricultural Economics Association meeting, Beijing, China, August 2009.

Petry, Mark, and Lei Zhang. *China's Corporate Income Tax Exemption for Agricultural Enterprises.* GAIN 9003. U .S. Department of Agriculture, Foreign Agricultural Service. January 13, 2009.

Ren, Xiaoming. "16 Questions on Treatment of Swine Fever (Zhu Gao Re Bing Shiliu Wenti)" (in Chinese), *Today's Animal Husbandry and Veterinary Medicine* (Jinri Xumu Shouyi), June 12, 2010.

Sun, De, Haiyan Jia, and Guochen Duan. "2011 Nian Zhu Jia Tui Shou Da Diao Cha (2011 Hog Price Survey) ." Breeding Hog Information Net, June 2011. http://www .chinaswine .org .cn/piaofu/dc201105/dcjg .htm

Sun, Luwei. "Sheng Zhude Ying Kui Ping Hengdian Shi Zenyang Huifude (How the Breakeven Point for Hogs Was Reached)," *Farmers Daily*, August 3, 2010. http://nc .people .com .cn/GB/12329988 .html

Van Arsdall, Roy N ., and Kenneth E. Nelson. *U.S. Hog Industry.* Agriculture Economics Report No. 511. U .S. Department of Agriculture, Economic Research Service, June 1984.

Wang, Yan'an. "*Yi Tou Zhude Jianyi zhi Lu* (A Pig's Quarantine Trip)," *Economic Observer News*, May 8, 2009.

Wang, Fanghao, Wenqi Ma, Zhengxia Dou, Lin Ma, Liu Xiaoli, Junxiang Xu, and Fusuo Zhang. "The Estimate of the Production Amount of Animal Manure and its Environmental Effect in China" (in Chinese), *China Environmental Science*, 2006, Vol. 26, No. 5, pp. 614-617.

Woolsey, M ., J. Zhang, and K. Rasmussen. *China Livestock and Products Annual Report.* GAIN CH10055, U .S. Department of Agriculture, September 24, 2010.

Woolsey, M ., and J. Zhang. *U.S. Pork Exports to China on the Rise.* GAIN CH11020, U .S. Department of Agriculture, May 9, 2011.

Xiao, Hongbo, and Jimin Wang. "China's Hog Production Situation in 2008 and Its Prospects" (in Chinese), *Agricultural Outlook*, June 2009, pp. 16-18.

Xinhua News Agency. "Fa Gai Wei: Zhongguo Bu Keneng Daliang Cong Waiguo Jinkou Zhu Rou (Development and Reform Commission: China Cannot Import Large Amounts of Pork from Foreign Countries)," online news report, September 4, 2007.

Yang, Wan. "Zhu Sheng Bing le, Bu Ru Zhijie Deng Si Na Baoxian? (Sick Pigs, Better to Wait For Them To Die to Collect Insurance?)," *Yangzhou Daily News,* July 1, 2010. www .yznews .com .cn/news/2010-07/01/ content_3278925 .htm

Zhang, Xuebiao, and Nie Fengying. "Retrospect and Prospect of the Chinese Pork Market" (in Chinese). *Agricultural Outlook*, March 2010, pp. 19-22.

Zhang, Yi. "Hog Farming: If Not the 'Roller Coaster' Which Car to Ride?" (in Chinese), *Peoples Daily*, June 27, 2010. http://nc .people .com .cn/ GB/11980484 .html

Zhang, Yue. "Xu Qin Yangzhi Ye Chengwe Nongye Wuran Yuan zhi Shou (Livestock and Poultry Industry Have Become the Leading Source of Agricultural Pollution)," Xinhua News Service, December 5, 2010. www. gov .cn/jrzg/2010-12/05/content_1759746 .htm

Zhou, Wangjun. "Positive Research on Stabilizing the Hog Market" (in Chinese). National Development and Reform Commission, May 2010. www .sdpc .gov .cn/zjgx/t20100525_348867 .htm

Zhu, Zhigang. "Zhu Zhigang Answers Questions on Response to Rising Pork Prices and Promotion of Healthy Hog Industry Development" (in Chinese). Ministry of Finance Net, June 22, 2007. www.gov.cn/ gzdt/2007- 06/22/content_657983 .htm

## End Notes

[1] We combine trade statistics for mainland China and Hong Kong since industry sources say that some pork imported to Hong Kong is re-exported to mainland China. Also, Hong Kong's pork imports shift away from mainland China to other suppliers when Chinese prices rise .

[2] According to Zhang (2010), the pork price dropped dramatically during the H1N1 (swine flu) crisis, but consumer confidence recovered rapidly after China's vice minister of agriculture made an announcement to reassure the public that pork was safe to eat .

[3] Chinese authorities also quarantined large numbers of travelers from North America during this period.

[4] Chinese feeder pig prices follow an even more pronounced cyclical pat-tern than do pork and live hog pricesFeeder pig prices rise during periods of industry expansion and fall when farms are cutting back on production.

[5] Data from the China Ministry of *Agriculture Livestock Industry Yearbook* indicate that approximately half of hogs slaughtered in China in 2010 came from "backyard" farms raising fewer than 50 hogs per year .

[6] Guangdong Province is a major hog-producing area in southern China, and it has the largest output of commercial animal feed of any province in China. Corn prices in Guangdong are among the highest in China because little corn is produced there. Several million metric tons of corn are shipped from northeastern China to Guangdong each year .

[7] The increase in China's consumer price index was 12 percent for all food and 4 .8 percent for all items during 2007 .

[8] A *Peoples Daily* article blamed the "herd mentality" and "synchronized behavior" of small farmers for market volatility and suggested that intervention cannot stabilize the market as long as small-scale farmers are predominant (Zhang, 2010a). The Ministry of Agriculture asserted that sow subsidies and improved market information prevented a large cull of sows during the period of low prices in 2010 (Sun, 2010) .

[9] Only one policy was aimed at pork consumers: small cash subsidies for low-income families and students from poor families to compensate for the rising cost of purchasing pork during a brief period in 2007. Temporary subsidies for general food-price increases were given in late 2010. During 2011, some local governments opened temporary shops selling discounted pork .

[10] These pork counties (*zhu rou chu diao xian*) are important suppliers of pork that produce more pork than they consume. They are chosen by ranking counties using an index based on the number of hogs sold outside the county, volume of hogs slaughtered, and inventory of hogs. The number of counties receiving awards increased each year during 2007-10 .

[11] Some of these local programs are financed by the financial awards and other subsidies included in the list above .

[12] A large cull of sows constrains the sector's ability to expand production in later months. This phenomenon was blamed for the sharp increase in pork prices in 2007 .

[13] Provincial guidelines specify that the ratio is to be calculated using the provincial average live hog price received by farmers and the wholesale price of corn. Secondary targets stated in the National Development and Reform Commission document are to prevent the ratio of feeder pig-to-hog carcass (*bai tiao rou*) prices from falling below 0 .7:1, maintain hog inventories of at least 410 million head, and maintain sow inventories of at least 41 million head .

[14] Provincial documents for implementing the program set 6:1 as the breakeven point for the hog-corn price ratio, but officials may adjust the targets based on changes in industry structure and other factors. Henan Province's document set the breakeven point at 7:1, and Zhejiang Province set it at 6 .5:1 .

[15] Zhejiang Province specified a 7-day supply of reserve pork for its three largest cities and a 3-day supply for other cities. As noted on page 19, the national required reserve for large cities was raised to a 10-day supply in 2011 .

[16] The means of limiting imports are not specified. The Shandong Province document explains that the provincial branch of the inspection and quarantine bureau will work with other departments to implement temporary policies to limit imports .

[17] This total may include modest amounts of other meats in addition to pork .

[18] Liu and Sun (2010) estimated that all sales of chilled, frozen, and processed pork together account for less than 10 percent of China's pork market.

[19] Woolsey et al. (2010) also report that low efficacy of free vaccines may contribute to epidemics. Some local distributors lack facilities to keep vaccines at the proper temperature.

[20] A company official told Liu that in some cases, multiple farmers filed claims on the same dead animal .

[21] Many supermarkets and specialty retail shops in China offer pork from local pig breeds. The appeal of this pork, generally sold at premium prices, is its stronger flavor and purported health benefits .

[22]Articles from official media included assurances that imported pork was purchased mainly by hotels, that imports would not affect the market, and that China has free trade in pork but also a strict inspection and quarantine system .

[23] This refers to China's high reliance on soybean imports .

In: Global Agriculture ISBN: 978-1-62618-260-8
Editors: M. R. Robertson, J. I. Schmidt 

*Chapter 3*

# LONG-TERM GROWTH PROSPECTS FOR WHEAT PRODUCTION IN AFGHANISTAN*

*Suresh Persaud*

## ABSTRACT

Afghanistan has had success at achieving higher output for wheat, its key staple food and major crop. From 1990-2009, wheat production grew 3.5 percent per year, driven primarily by yield increases. However, given expected increases in demand, imports are likely to grow in coming years even if this production growth is sustained, suggesting growing dependence on supplies from Pakistan and other countries. For Afghanistan to achieve self-sufficiency in the next decade, more widespread use of improved seed and fertilizer on irrigated and rainfed wheat fields would have to be combined with sharp increases in irrigated wheat area—an unlikely outcome.

## INTRODUCTION

Afghanistan, strategically positioned between Central Asia, the Middle East, and South Asia, is a landlocked country where key threats to food security stem from weak transportation links that limit Afghanistan's access to

* This is an edited, reformatted and augmented version of United States Department of Agriculture, Publication No. WHS-11L-01, dated January 2012.

international markets, coupled with a domestic agricultural sector that is inadequate for meeting food needs. In Afghanistan, wheat is a staple food, accounting for over half of the population's caloric intake on average (Chabot and Dorosh, 2007). Imports from neighboring countries have played a key role in stabilizing wheat and flour prices in Afghanistan. Although Afghanistan imports wheat from a number of neighboring countries, Pakistan has historically supplied more than half of these imports.

Afghanistan's location and heavy reliance on Pakistan for imported wheat leaves Afghanistan vulnerable to fluctuations in Pakistani supplies and trade policies. For example, in 2008 Afghanistan's food supply network broke down due to a confluence of events, including shortfalls in Pakistani and Afghan wheat production and Pakistan's bans on wheat and flour exports (Persaud, 2010). Given Afghanistan's landlocked position, its reliance on a small number of regional suppliers to meet import needs, and continuing uncertainties surrounding the reliability of these imports, this study focuses on the long-term prospects for expanding domestic wheat production.

**Table 1. Wheat area, yield, and production growth in Afghanistan**

| Period average | Area | Yield[1] | Production |
|---|---|---|---|
| | *(1,000 ha)* | *(mt/ha)* | *(1,000 mt)* |
| 1979-81 | 2,300 | 1.20 | 2,754 |
| 1989-91 | 1,623 | 1.06 | 1,725 |
| 2008-10 | 2,355 | 1.42 | 3,350 |
| | | *Growth rates (percent)*[2] | |
| 1980-90 | -4.10 | -1.30 | -5.3 |
| 1990-2009 | 1.51 | 1.97 | 3.5 |
| 1980-2009 | 0.10 | 0.40 | 0.5 |
| mt = metric ton. ha = hectare. | | | |

[1]Computed by dividing 3-year average of production by its respective 3-year average of area. [2]Least squares growth rate.

Source: USDA, Foreign Agricultural Service, Production, Supply and Distribution (PSD) online database, and author's calculations.

Afghanistan has had some success at achieving higher wheat output. From 1990-2009 wheat production grew 3.5 percent per year, driven primarily by yield increases that reflect long-term agricultural assistance efforts by international organizations and the presence of functioning seed and fertilizer markets (table 1) (Maletta, 2007; Favre, 2004; Maletta and Favre, 2003).

However, model-based simulations indicate that this production growth, even if sustained until 2020, will not outpace expected increases in consumption, suggesting growing dependence on supplies from Pakistan and other countries. To achieve self-sufficiency in the next decade, more widespread use of improved seed and fertilizer on irrigated and rain fed wheat fields would have to be combined with improbably sharp increases in irrigated wheat area.

## Growth and Instability of Wheat Production

With mountainous terrain comprising roughly three-quarters of Afghanistan's land area and an arid to semi-arid climate, arable land in 2008 amounted to 7.794 million hectares or approximately 12 percent of the country's area (FAO, 2011). Most of Afghanistan's agricultural and irrigation development occurs along the country's few fertile valleys, which are formed by rivers extending from the Suleiman and Karakoram mountains and the Hindu Kush range (Rout, 2008). The vast majority of Afghanistan's wheat area and production is located in the northern plains bordering Turkmenistan and Uzbekistan (USDA, 2008).

During the main growing period there is little, if any, reliable rainfall, meaning that Afghanistan must depend on irrigation to meet the majority of its crop water requirements. Winter snowfall in the mountain ranges of central Afghanistan supplies over 80 percent of the country's annual precipitation (USDA, 2008). The primary storehouse of the country's irrigation water is in the Hindu Kush range (Rout, 2008). Snowmelt in the spring is the major source of irrigation water, running through rivers and streams that originate in the mountains. Given the absence of sufficient rainfall during the critical growing period, the timing and duration of annual snowmelt is a key factor in determining the volume of irrigation water and the length of time that it is available (USDA, 2008).

### Uneven Historical Trends

In 1976 Afghanistan harvested 2.94 million metric tons (mmt) of wheat from about 2.4 million hectares (ha) of land, implying a yield of 1.23 mt/ ha. Production remained stable for about 10 years thereafter, and then deteriorated sharply beginning in 1987 (figure 1). Yield and area planted to wheat contracted, and by 1992 Afghanistan's wheat output was little more than half of its 1976 level (figure 2). Grain production deteriorated due to the combined effects of prolonged droughts and the destruction of irrigation networks and infrastructure during years of conflict, first between Afghan and Soviet forces and then during the civil war that followed the 1989 Soviet withdrawal.

Wheat production trended upward from 1993-99, driven primarily by strong gains in yield. At 1.3 mt/ha in 1998, yield was at a then historic high in Afghanistan. However, throughout the 1993-99 period, the quantity of wheat harvested never exceeded the 1976 peak due to relatively slow growth in area cultivated to wheat. Moreover, these production gains proved to be fragile. Successive droughts in subsequent years led to substantial crop shortfalls in irrigated as well as rain-dependent areas, particularly in 2000 and 2001 when wheat production reached new lows, falling below the 1992 level.

Growing conditions were relatively favorable between 2002 and 2007. In 2003 production finally surpassed the earlier 1976 peak. The 2003 wheat harvest was estimated to be 3.55 million metric tons, a level that Afghanistan achieved again in 2005 and approached in 2007. These production peaks were achieved primarily as a consequence of higher yields. Throughout the 2002-07 period, there was no year in which estimated area planted to wheat exceeded the mid-1970s level. Yield, on the other hand, reached a new high in 2007 at 1.52 mt/ha, which was significantly above the previous 1998 peak (1.3 mt/ha).

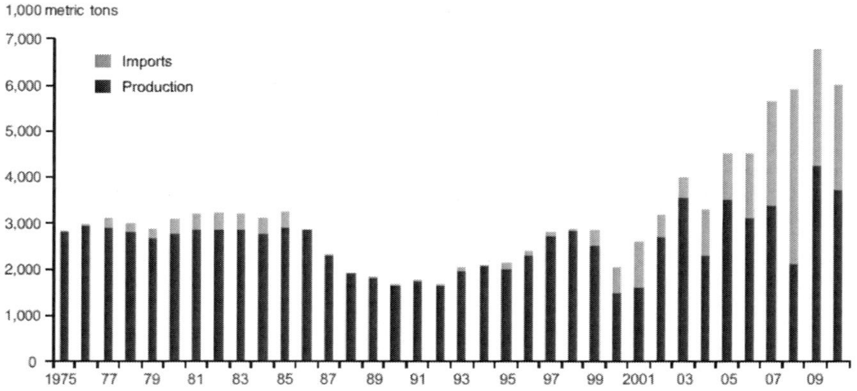


Notes: Includes flour imports expressed in wheat equivalent. One ton of wheat converts, with some variation, to approximately 0.74 tons of flour. Data are on a marketing-year basis; 2008 refers to July 2008–June 2009.

Source: USDA, Foreign Agricultural Service, Production, Supply and Distribution (PSD) online database.

Figure 1. Wheat and flour supply in Afghanistan.

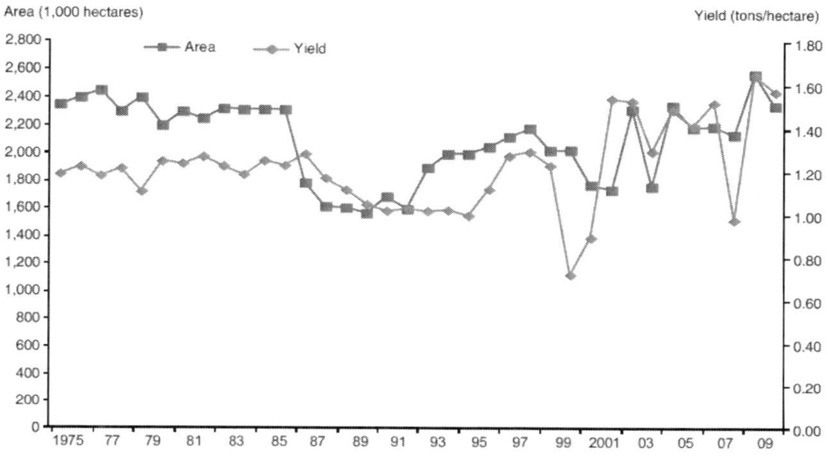


Notes: Data are on a marketing-year basis; 2008 refers to July 2008–June 2009.
Source: USDA, Foreign Agricultural Service, Production, Supply and Distribution (PSD) online database.

Figure 2. Wheat area and yield in Afghanistan.

The years of favorable wheat harvests were followed by below-average levels of rain and snow during the 2007/08 wet season, which reduced Afghanistan's 2008/09 wheat crop to 2.1 million metric tons.

Excellent growing conditions allowed Afghanistan's wheat crop to rebound in 2009/10 to a record 4.3 million tons. Yield rose to a new high of 1.65 tons per hectare, a 68-percent increase over the previous year, and significantly above the previous 2007 peak (1.52 mt/ha). At 2.6 million hectares in 2009/10, area planted to wheat finally rose above the 1976 level. Growing conditions have continued to be favorable and USDA forecasts that 2010/11 wheat production and yield will be 3.7 million tons and 1.57 tons per hectare, respectively (USDA, 2011).

## Efforts to Raise Wheat Yields

The growth in wheat yields reflects long-term efforts at seed development and availability in Afghanistan supported primarily by the United States Agency for International Development (USAID), the Food and Agriculture Organization of the United Nations (FAO), and the International Center for Agricultural Research in the Dry Areas (ICARDA). These yield increases also demonstrate that seed and fertilizer markets have continued to exist and function, albeit with significant disruptions and under a great deal of adversity (Maletta, 2007; Favre, 2004; Maletta and Favre, 2003). This has allowed yield to reach a number of new highs when favorable growing conditions occur.

International organizations developed contracting arrangements with Afghan farmers to achieve the adoption of higher yielding seed varieties. These arrangements created demonstration-effects, allowing neighboring farmers to observe and gauge the impacts of adopting new seed varieties, while also facilitating farmer-to-farmer exchanges (Favre, 2004). These demonstration effects likely encouraged the adoption of higher yielding seeds through market-based mechanisms and supply chains.

Available information from the 2001-02 season indicates that wheat farmers in irrigated areas relied to a significant degree on markets to obtain seed (Maletta and Favre, 2003). In 2003, 58 percent of the seed planted was the farmers' own (saved) seed[1]; approximately 40 percent was purchased in local bazaars, and the remainder was obtained from other sources such as humanitarian organizations. Similar results were found in rainfed wheat areas, where the farmers' own (saved) seed and bazaars provided 43 percent and 53 percent, respectively, of the wheat seed planted. In the irrigated wheat subsector, the data show a replacement rate of 39 percent for seed, which suggests a willingness on the part of Afghan wheat farmers to take risks to raise yield and hence production through a process of rapid adoption of improved seed (Favre, 2004).

Total fertilizer use on all crops just before the Soviet invasion in 1979 is estimated at only about 50,000 mt (or roughly 20 kgs/ha). In recent years, however, the use of fertilizer has become widespread in irrigated wheat areas, with a national average of 179 kg per fertilized hectare of irrigated wheat in 2001-02. The 78 percent of surveyed farmers who used fertilizer on irrigated wheat, typically consisting of urea in the spring and diammonium phosphate (DAP) in the autumn, applied a total of 187,000 mt. This growth in nutrient application was fostered by private traders who sell imported fertilizer throughout the country in the absence of Government programs. The growth underscores the key role of markets in supplying agricultural inputs, despite chaotic, war-torn conditions (Favre, 2004). Fertilizer use continued to grow, by almost 50 percent, to 300,783 mt in 2003 (Roy, 2003).

The fertilizers used in Afghanistan are mainly urea and DAP, and to a lesser extent mono ammonium phosphate obtained from Iran. Afghanistan's demand for DAP is met through imports from the United States, Australia, China, and Pakistan. Fertilizer from the United States and Australia tends to be shipped first to Pakistan, where it is repackaged and then sold to Afghanistan. Although Afghanistan imports urea from a number of countries, including Tajikistan, Iran, Uzbekistan and Saudi Arabia, the major supplier is Pakistan. Afghanistan's relatively high prices attract shipments of urea and DAP from Pakistan. The Government of Pakistan's fertilizer subsidies, including those that seek to cap domestic Pakistani prices, may be a factor contributing to cross border movements (Husain and Gilroy, 2009).

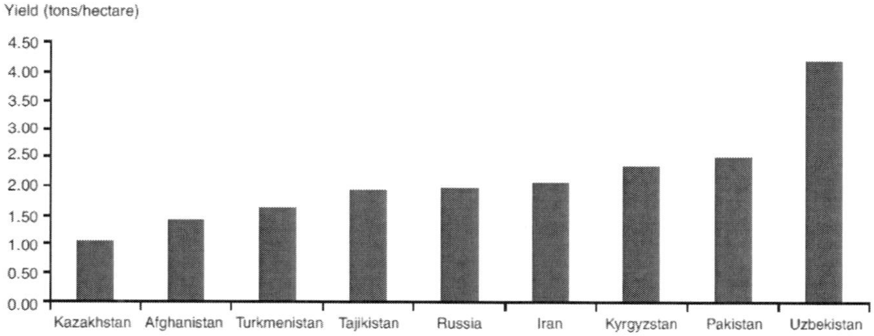


Sources: USDA, Foreign Agricultural Service, Production, Supply and Distribution (PSD) online database.

Figure 3. Wheat yields in Afghanistan and in neighboring countries (2002-09 average).

Although functioning markets for seed and fertilizer have allowed growth in yield and output, at 1.4 mt/ha (2002-09 average), Afghanistan's yield is still low relative to most other countries in the region (figure 3). Moreover, the historical performance of Afghanistan's wheat production sector reveals its sensitivity to variations in weather and precipitation. The destruction and continuing disrepair of the country's irrigation systems, aside from exacerbating the impact of adverse growing conditions, likely also explain the negligible growth in wheat area, which in turn constrained growth in output. Consequently, imports of wheat and flour have played a key role in fostering the growth and stability of Afghan consumption.

## Consumption and Imports

Household surveys demonstrate the centrality of wheat in Afghan food consumption patterns, as wheat accounts for approximately 60 percent of total caloric intake (Government of Afghanistan, 2003). Indeed, the majority of the poor in Afghanistan rely largely on fresh bread and green tea for breakfast and lunch. At 186 kilograms per person per year (2002-09 average), per capita consumption of wheat in Afghanistan was below neighboring countries such as Turkmenistan, Uzbekistan, and Iran but significantly above Pakistan (figure 4).[2] However, considerable uncertainty surrounds estimates of Afghanistan's population and consumption, implying that per capita consumption figures are likewise uncertain (see appendix 3).

Apparent per capita consumption of wheat has grown rapidly in recent years, increasing 7.4 percent per year from 2002-09 despite double-digit growth (11 percent) in real flour prices (figure 5).[3] Afghanistan's economy expanded rapidly over this period—official real (adjusted for inflation) gross domestic product per capita grew at an annual rate of 7.3 percent (IMF, 2010).[4] Growth in wheat consumption cannot be attributed completely to rising income. Rehabilitation of the country's infrastructure and marketing channels allowing increased flows of grain into remote areas, combined with a relatively stable macroeconomic environment, have also fostered an upward trend in wheat consumption.[5] Rising incomes are expected to lead to dietary diversification in Afghanistan (Maletta, 2007), and current levels of consumption are already high. Future growth in consumption may not be as rapid as in the past decade.

## Role of Wheat Trade

Given the wide fluctuations in Afghanistan's wheat production, imports have helped to stabilize consumption levels and meet the food needs of Afghanistan's growing population (see figure 1). Based on a measure of temporal variability known as the coefficient of variation, Afghan production has been three and a half times more variable than consumption, as imports have partially offset poor domestic harvests due to droughts, thereby dampening fluctuations in Afghan consumption (Persaud, 2010).

Afghanistan is a landlocked country, bordered on the west by Iran, on the south and the east by Pakistan, and on the north by Turkmenistan, Uzbekistan, and Tajikistan. Afghanistan imports wheat and flour from a number of neighboring countries. However, with a shared border of 1,600 kilometers and a long history of trade, Pakistan is the dominant supplier of wheat to Afghanistan. The private millers and traders of Pakistan work closely with Afghan traders (Khan, 2007).

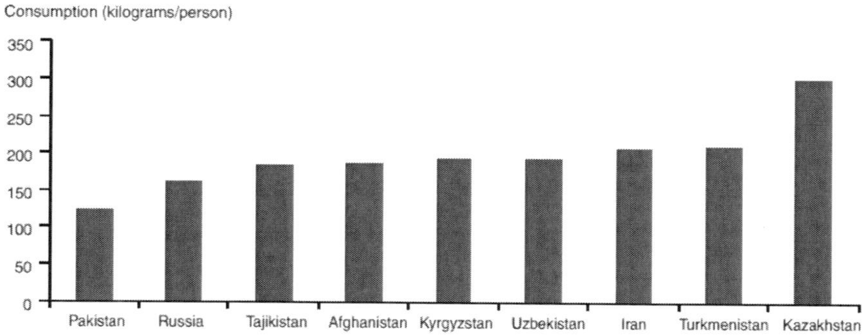


Sources: USDA, Foreign Agricultural Service, Production, Supply and Distribution (PSD) online database; and UN FAOSTAT database.

Figure 4. Wheat consumption in Afghanistan and in neighboring countries (2002-09, average).

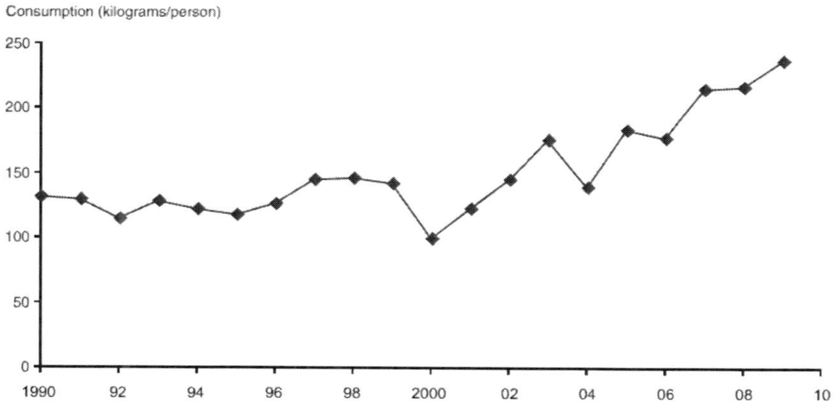


Sources: USDA, Foreign Agricultural Service, Production, Supply and Distribution (PSD) online database; and UN FAOSTAT database.

Figure 5. Per capita consumption of wheat in Afghanistan.

Pakistani wheat flour is widely accepted by Afghan consumers because of its quality, and Pakistani mills extend credit to Afghan traders seeking to purchase flour (Schulte, 2007; Khan, 2007).

## Links to Pakistan Market

The Government of Pakistan has implemented price policies aimed at improving the availability of food for its population (Dorosh and Salam, 2006). The Government intervenes in domestic wheat markets by procuring wheat from farmers at a support price. The Government also releases wheat to the provinces, which is sold to the flour mills at a Government-determined "issue" or "release" price (USDA, 2009). These interventions, which tend to involve sales of wheat to flour mills at below-market rates, generate profits for millers while also incurring fiscal costs for the Government because the issue prices do not cover the full cost of procurement (domestic or imported), storage, and handling (USDA, 2009; Dorosh and Salam, 2006). The growth of Pakistan's flour mills has been concentrated in provinces neighboring Afghanistan, and a large number of Pakistani mills operate only when they are able to purchase subsidized wheat from their Government (Khan, 2007). Pakistani traders have incentives to move flour into relatively high-priced nearby markets, thus providing strong competition for Afghan millers.

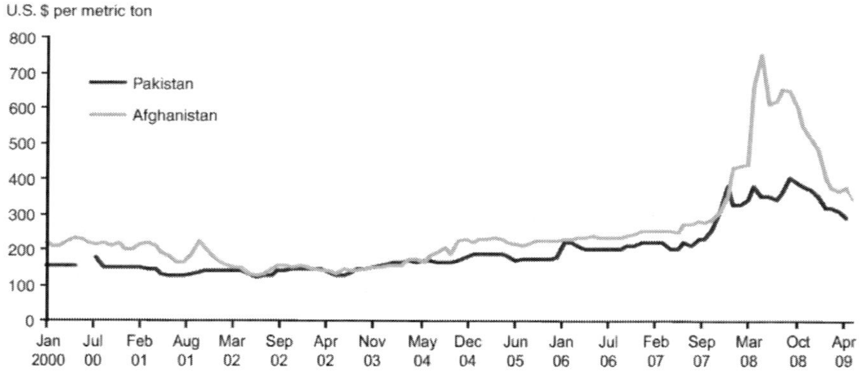


Figure 6. Prices of wheat in Afghanistan and Pakistan.

Since Pakistan is a major player in the Afghan grain market, wheat prices in Pakistan and Afghanistan tend to move in the same direction. Using monthly data for January 2002 through June 2005, Chabot and Dorosh (2007) conducted formal econometric tests to explore issues of (1) market integration between major markets within Afghanistan, and (2) market integration between Pakistani and Afghan markets. Cointegration results suggest that wheat prices in major markets in Afghanistan and in Lahore, Pakistan, tend to move together in the long run (Chabot and Dorosh, 2007). Changes in Pakistan's Government-fixed release or issue prices, to the extent that they influence market prices of wheat within the Pakistani market, would also shape prices that prevail in Afghanistan. Indeed, from 1999-2009, the real issue price of wheat in Pakistan and the real retail price of wheat in Afghanistan are well-correlated (correlation coefficient = 0.91), implying that shifts in Pakistan's price policies impact Afghan consumers.

In addition to Pakistan's price policies, its trade policies and domestic supply situation have influenced Afghan wheat prices. Although wheat price changes in these two countries have, over the years, generally been similar in terms of directionality, the gap between Afghan and Pakistani wheat prices increased sharply in 2008 (figure 6). Afghanistan's food supply network broke down due to a confluence of events, including shortfalls in Pakistani and Afghan wheat production and Pakistan's bans on wheat and flour exports (Persaud, 2010).

As supply disruptions eased, Afghan-Pakistani price gaps decreased after May 2008. More recently, Pakistan experienced large-scale flooding after unusually heavy monsoon rains that began at the end of July 2010. This latest shock to Pakistani agriculture is not an immediate threat to Afghan food security due to a combination of relatively abundant Afghan supplies, the availability of imports from Kazakhstan, and notably, Pakistan's high levels of wheat stocks that allow Pakistan to continue exporting. Also, despite the severe flooding, Pakistan's 2010/11 wheat crop is estimated at 23.9 mmt, only slightly below the previous year's (2009/10) record wheat production of nearly 24.0 mmt. In December 2010 the Government of Pakistan removed its ban on exports, allowing private traders in the country to export up to 1.0 mmt of wheat. The decision to lift the export ban will likely result in increased exports to Afghanistan (USDA, 2010).

Significant future growth in Afghan wheat consumption is likely, if domestic production and Pakistani policies support such growth. We expect population to grow at an annual rate of 2.25 percent through 2020, in line with Maletta (2007). Real GDP is projected to grow 6.94 percent per year from 2009-14, consistent with a recent International Monetary Fund Country Report (IMF, 2010).

Population and income growth rates of these magnitudes raise the question of whether domestic production of wheat will keep pace with consumption, or whether Afghanistan will experience a widening gap between demand and domestic supply.

# Prospects for Afghanistan's Wheat Market

Afghanistan's agricultural potential is limited by its predominantly mountainous terrain and arid to semi-arid climate. Nevertheless, further increases in wheat production and productivity are possible.

## Water and Irrigation

Afghanistan's water resources include an estimated 75 billion cubic meters (bcm) of potentially available renewable water resources annually, comprised of 57 bcm of surface water and 18 bcm of ground water. Given the country's low precipitation, ground water supplies are recharged mainly by surface water. Although agriculture is the primary user of water, accounting for approximately 93 percent of Afghanistan's actual total water use, irrigation consumes only about 20 bcm per year—far less than the country's potentially available annual renewable water resources (75 bcm) (Ahmad and Wasiq, 2004). The U.N. Food and Agriculture Organization estimates that 7.794 million hectares of Afghan land were arable in 2008 (FAO, 2011). However, given the country's relatively low and variable rainfall, irrigation is critical—approximately 85 percent of the total agricultural output and 70 percent of wheat production are supplied by irrigated areas. Afghanistan's potential irrigable area is about 5 million hectares, well above the currently developed area of about 2.6 million hectares (ADB, 2003). The relatively modern, formal river-diversion structures that were built in the 1960s and 1970s in Afghanistan account for only 15 percent of the total irrigated area. Traditional systems that employ informal river-diversion structures maintained by the users account for about 55 percent of the total irrigated area, while traditional systems that are based on natural springs account for the remaining 30 percent. The very low efficiency rating of these irrigation systems (about 25 percent) indicates substantial room for reducing water wastage and increasing the irrigated area.[6] Aside from damage resulting from conflicts and lack of maintenance, salinity and water logging problems affect a number of the more modern systems and larger traditional systems (ADB, 2003). Mainly because of unpredictable water availability that has been exacerbated by years of military conflict, in 2009 only 3.2 million hectares, or less than half of the land that could be used for crops, was actually cultivated. Since only 3 of Afghanistan's 34 provinces have fully operating irrigation systems in place (Torell and Ward, 2010), cropped area may grow as a consequence of investments aimed at rehabilitating existing irrigation networks, without the necessity of building new dams (Maletta, 2007). Additions to cropped area, being irrigated land, would be expected to provide relatively high yields. However, efforts to exploit water resources may lead to conflict among Afghan communities. Water is a shared resource, and projects that are designed to harvest water in one location may reduce its availability in adjacent farming communities. In addition to the lack of technical and financial resources, the issue of water rights has been a complicating factor in the development of Afghanistan's surface-water irrigation systems (Rout, 2008). Land has remained idle, not only as a consequence of physical damage to canals, but also as a result of shortcomings in water management, depriving some farmers of access to water (Maletta, 2007). Moreover, programs to harness Afghanistan's rivers may also generate tensions with neighboring countries, to the extent that such investments reduce the flow of water to adjacent countries that share the same river basin with Afghanistan. Afghanistan's three major watersheds include (1) the Helmand River, flowing from the southern slopes of the Hindu Kush and proceeding southwest to the Sistan Basin in Iran; (2) the Kabul River, which begins in the southeastern Hindu Kush and moves south through the city of Kabul, before turning east and joining the Indus River in Pakistan; (3) the Amu Darya, which begins on the northern slopes of the Hindu Kush. The Amu Darya basin is notable in that it extends into the Kyrgyz Republic, Tajikistan, Turkmenistan, and Uzbekistan, as well as Afghanistan (Ahmad and Wasiq, 2004). Large-scale investments aimed at actualizing Afghanistan's irrigation potential may in some cases require treaties with affected downstream countries.

Although farmers have responded to limited access to surface water by digging wells, ground water is a critical, and, for many households, the sole source of drinking water. Water tables in Afghanistan are already falling, requiring deeper wells. Shifts toward greater agricultural usage of ground water is risky and will likely come at the cost of human consumption (United Nations, 2008). Given the difficulties and the potential for conflict that are associated with expanding the country's irrigation sector, imports of wheat have played an important role in allowing Afghanistan to compensate for its own domestic constraints on water availability. Water, a fundamental input for crop production, is embedded in agricultural products—trade in agricultural commodities can be viewed as a virtual flow of water from producing and exporting countries to importing countries. Thus, arid and semi-arid countries can achieve welfare improvements through food imports, while reducing potential competition among neighboring countries that share the same river basins.

## Yield Potential

Observations of farmers who have adopted improved inputs, as well as evaluations of emergency programs that distribute fertilizer and improved seed, provide a rough indicator of the potential for increasing wheat productivity in Afghanistan.

Almost 20 years ago, the Swedish Committee Agricultural Survey estimated that yields from improved seed varieties ranged from 2.98 mt/ha in the Northwest to 3.47 mt/ha in the East Central region in 1993 (Swedish Committee for Afghanistan, 1993). A key point is that these were actual observed yields obtained by farmers and not experimental or expected yields. Since 1993, newer varieties of improved seed with even higher yields have been introduced. In 2001-02 and 2002-03 improved seeds yielded 3.0-4.0 mt/ha on average. Depending largely on soil quality as well as factors such as the availability and use of water and fertilizer, many Afghan farmers achieved yields that were substantially higher than the average, e.g., yields in the range of 5.0-5.75 mt/ha (Maletta and Favre, 2003).

FAO distributed improved seed as part of its Agricultural Emergency program for the 2002-03 agricultural year and the result was an average yield of 3.30 mt/ha among program beneficiaries, who were mostly poor small farmers (Mollet, 2003).

Operating through nongovernmental organization (NGO) partners, the International Center for Soil Fertility and Agricultural Development (IFDC) and the International Center for Agricultural Research in the Dry Areas (ICARDA) provided fertilizer and improved seed to Afghan wheat farmers in an emergency distribution program. This program, which reached approximately 200,000 farmers in 13 provinces, entailed the distribution of about 15,700 mt of fertilizer for wheat in the spring and fall of 2002 and in the spring and summer of 2003. A combination of favorable weather conditions, the availability of fertilizers (mainly urea and DAP), and improved seed resulted in average wheat yields of 4 mt/ha with a considerable degree of variability around the average.

Estimates of incremental yield indicate that 15,700 tons of fertilizers provided an additional yield of 78,500 tons of wheat, i.e., 1 kg of fertilizer was associated with an additional 5 kg of wheat (Gregory, 2006). A great deal of caution is warranted if the objective were to generalize this incremental yield estimate beyond the recipient farmers and beyond the window of time in which the program was implemented. The estimated relationship between additional fertilizer and yield gain is conditional upon factors such as soil quality, weather conditions, and farmers' skill and knowledge. However, this fertilizer and seed distribution program did cover 13 agriculturally important provinces, and the average realized yield of 4 MT/HA is consistent with results noted earlier in this report (Gregory, 2006).

Oushy (2010) suggests that wheat yield can be increased by improving and diversifying crop rotations beyond the typical wheat/rice or wheat/corn cycle, thereby sustaining soil properties and productivity. These commonly used crop rotations have resulted in deteriorating soil condition and proprieties, including lower fertility, diminished organic matter, lower soil water-holding capacity, salinity, and weed and disease invasion (Oushy, 2010).

Moreover, since less than half of Afghanistan's wheat area is sown to high-yielding varieties, the average yield for the country as a whole will trend higher as the use of superior seed becomes more widespread. This will entail exploiting the genetic potential of existing seeds that are already available for use, rather than developing different strains. When new seed varieties were used in 2002-03, average wheat yields for irrigated fields were about 3 mt/ ha (peaking at more than 6 mt/ha in some locations), while for rainfed areas the average was 1.1 mt/ha with peaks up to 3 mt/ha. These figures could understate yield potential—it is important to note that seeds classified as higher yielding include older strains as well as improved seeds that exhibit declining yield as a consequence of having been genetically contaminated by inbreeding with other varieties.

## Projections to 2020

In the longer term, domestic wheat production in Afghanistan could expand significantly with greater use of improved inputs. However, even with dietary diversification tempering demand growth, wheat consumption could also grow significantly, propelled by strong growth in population, as well as improvements in marketing chains that connect consumers to domestic and external supplies of food.

It is not possible to project the prospects for Afghanistan's wheat market with certainty, given the impossibility of accurately forecasting the future growth of income, population, and food prices, and the unpredictable pace of improvements in technology and irrigation. Accordingly, we develop a quantitative framework that is capable of analyzing a range of different time paths and/or growth rates of key drivers of supply and demand. Specifically, we evaluate the likely role of yield growth and area expansion in the future growth in supply, and in meeting Afghan wheat demand using a multiyear simulation model (see box, "Characteristics of the Simulation Model").

### *Overview of Scenarios*

Based on different scenarios of future growth of Afghanistan's wheat yields, we develop projections through 2020 of Afghanistan's domestic wheat production, imports, and consumption. The three scenarios considered in the model are similar in a number of ways. In all scenarios, for the 2010-20 period, real GDP grows at an annual rate of 6.0 percent (table 2). (By way of comparison, *IMF Country Report* (2010) predicts that real GDP will grow in the range of 6 to 7 percent per year for 2009-15.) Afghanistan's population grows at annual rate of 2.3 percent as in Maletta (2007). Stocks of wheat and flour are assumed to be negligible. The Afghan-Pakistani border remains porous, and moreover, neither Afghan nor Pakistani trade policies hinder wheat trade, implying that trade is based on Afghan demand and relative prices. Throughout the projection period, the prices of wheat and flour in Afghanistan track Pakistan's release price, which in turn moves in lockstep with projections of Pakistan's government producer price generated by the USDA Baseline (USDA, 2011). Afghan wheat production does not rise rapidly enough to exert downward pressure on domestic flour prices in any scenario, i.e., Afghan flour prices do not fall below the margin-adjusted Pakistani release price. Since consumers face an identical price vector in all scenarios and income growth is the same in all cases, projected wheat consumption does not differ from scenario to scenario.

**Characteristics of the Simulation Model**

The analytical framework incorporates behavioral relationships for consumption and area cultivated to wheat. Farm production of wheat in Afghanistan is an identity, computed as the product of area and yield. Similar to the USDA Baseline models, wheat area is specified as an increasing function of the expected returns from its cultivation, where the lagged wheat price and lagged yield represent expected returns.

Yield growth is driven by various trend terms that represent differing rates of improvements in technology and irrigation.

Consumption of flour (on a wheat-equivalent basis) is a function of income, population, and the own-price, i.e., the price of flour. Stocks are assumed to be zero. Imports are computed by subtracting domestic production from consumption, where all variables are on a wheat-equivalent basis. Pakistan's release price of wheat influences both wheat and flour prices in Afghanistan.

A more complete description of the model, including the model parameters, is given in appendix 1 and the equations are provided in appendix 2.

**Table 2. Major Afghan wheat model analytical assumptions**

| Variable | Reference | High yield | Self-sufficiency |
|---|---|---|---|
| | | Percent | |
| Population growth | 2.25 | 2.25 | 2.25 |
| Real Gross Domestic Product growth | 6.00 | 6.00 | 6.00 |
| Yield growth | 0.90 | 1.97 | 5.50 |

Source: USDA, Economic Research Service.

The critical distinctions among the three scenarios are that they incorporate different assumed rates of yield growth. In the Reference case (Scenario I), yield grows at an annual rate of 0.9 percent, which matches the projected yield growth for Other Asia & Oceania generated by the USDA Baseline (USDA, 2011).[7] For perspective, this 0.9 percent rate of increase is above Afghanistan's 1980-2009 annual yield growth of 0.4 percent shown in table 1, but less than half of the country's more recent 1990-2009 annual yield growth of 1.97 percent. In Scenario II, yield continues to increase at its post-1990 rate of 1.97 percent per year to 1.91 mt/ha in 2020. Scenario III, for illustrative purposes, indicates the types of production and productivity advances that would likely be necessary for Afghanistan to eliminate its reliance on imports.

The model replicates the base year levels of Afghan wheat area, production, imports, consumption, and prices. The base year for the projections is 2010. The prior year is not an appropriate choice for the base period because the 2009 figure for apparent per capita consumption of wheat is an outlier— apparent consumption per person spiked to 237 kgs per person in

2009 (figure 7). The consumption data are uncertain, as data on stocks are not available. Nevertheless, an advantage of using a 2010 base is that consumption for that year is more consistent with the recent historical trend, whereas a 2009 base amounts to starting the projections from an extreme observation. However, in evaluating the projections, it is changes relative to the base year—as well as the relative differences between the various scenarios—that are relevant. Moreover, the scenarios presented herein are neither forecasts nor predictions.

***Scenario I: Reference***

Afghan domestic prices of wheat and flour, which are shaped by Pakistan's government-determined release price, exhibit almost no growth in real terms from 2010-20 (table 3). On a wheat equivalent basis, per capita flour demand (consumption per person) grows 1.2 percent per year over the same period, due to a combination of growth in per capita GDP of 3.7 percent per year and stability in flour prices.

Total demand, as distinct from per capita consumption, increases at an annual rate of 3.5 percent to 8.4 mmt in 2020. Hence, projected increases in consumption are not as rapid as in the past decade (figure 8). Nevertheless, this growth in demand still exceeds projected increases in output, as indicated below.

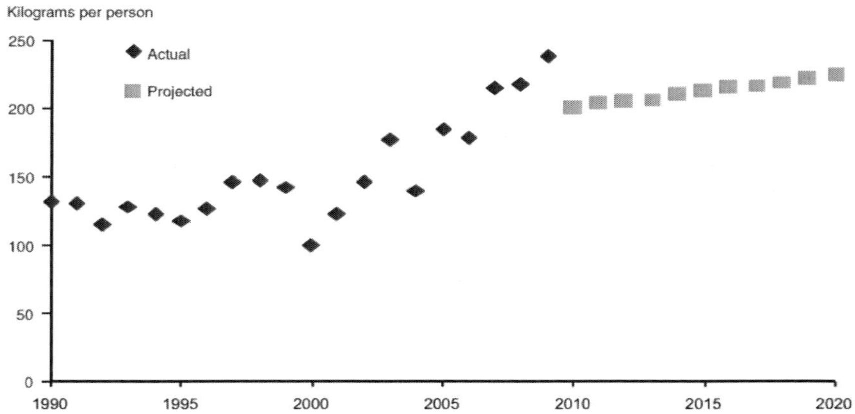


Sources: 1990-2010 data are from USDA, Foreign Agricultural Service, Production, Supply and Distribution (PSD) online database.
ERS model replicates 2010 data; 2011-20 figures are results of ERS model.

Figure 7. Projected per capita consumption of wheat (all scenarios).

**Table 3. Results of Afghanistan wheat model**

| | Scenario I Scenario IIScenario III | | | | | | |
|---|---|---|---|---|---|---|---|
| | Reference | | | High yield | | Self-sufficiency | |
| | Base year | Terminal year | Annual growth | Terminal year | Annual growth | Terminal year | Annual growth |
| | | | *Percent* | | *Percent* | | *Percent* |
| Area (1,000 ha) | 2,350 | 2,666 | 1.3 | 2,774 | 1.7 | 3,147 | 3.0 |
| Yield (mt/ha) | 1.57 | 1.72 | 0.9 | 1.91 | 1.97 | 2.68 | 5.5 |
| Production (1,000 mt) | 3,700 | 4,591 | 2.2 | 5,308 | 3.7 | 8,427 | 8.6 |
| Consumption (1,000 mt) | 6,000 | 8,427 | 3.5 | 8,427 | 3.5 | 8,427 | 3.5 |
| Per capita consumption (kgs/person) | 199 | 224 | 1.2 | 224 | 1.2 | 224 | 1.2 |
| Wheat imports (1,000 mt) | 2,300 | 3,836 | 5.2 | 3,119 | 3.1 | 0 | -- |
| Afghan wheat price ($/mt) | 341 | 340 | 0.0 | 340 | 0.0 | 340 | 0.0 |
| Afghan flour price ($/mt) | 498 | 497 | 0.0 | 497 | 0.0 | 497 | 0.0 |
| GDP ($ millions) | 15,608 | 27,952 | 6.0 | 27,952 | 6.0 | 27,952 | 6.0 |
| Population (millions) | 30.2 | 37.7 | 2.3 | 37.7 | 2.3 | 37.7 | 2.3 |

mt = metric ton.
ha = hectare.
kg = kilogram.
GDP = Gross Domestic Product.
Sources: Area, yield, production, consumption, and imports are from USDA, Foreign Agricultural Service, Production, Supply and Distribution online database. Population and GDP are from International Monetary Fund (2011). Prices of wheat and flour are from the U.N. World Food Programme's Price Analysis in Afghanistan (World Food Programme, 2011).

Table 4 shows the average yields on irrigated regions (2.36 mt/ha) and rainfed regions (0.92 mt/ha) for the base period. Yield on irrigated land is 2.6 times higher than on rainfed areas.

This figure, which matches the irrigatedto-rainfed ratio computed from yield data provided by Maletta (2004) for 2002/03, is also similar to the yield ratio computed from figures recently provided by CIMMYT.[8] Also in the base period, the proportion of irrigated (rainfed) area is 45 percent (55 percent). The ratios of irrigated-to-rainfed wheat areas and irrigated-to-rainfed wheat yields result in a national average yield of 1.57 mt/ha for Afghanistan as a whole, which matches the 2010 data from USDA (2011).

Farmers rely on functioning seed and fertilizer markets to expand the share of land cultivated with improved seed varieties and to increase nutrient application. In the Reference scenario, the yield in Afghanistan, i.e., the national average, grows at an annual rate of 0.9 percent, from 1.57 mt/ha in the base period to 1.72 mt/ha in the terminal year (figure 9).

The yield increases obtained in the Reference scenario do not require that Afghanistan increase the proportion of its irrigated wheat area or that the country improve its yields on rainfed areas. By assumption, the rainfed yield remains constant at 0.9 mt/ ha throughout the projection period, while in contrast the irrigated yield rises from 2.36 mt/ha in 2010 to 2.7 mt/ha in 2020 (see table 4). If the proportion of irrigated wheat area remains at 45 percent throughout the projection period and the ratio of irrigated-to-rainfed yields climbs to 2.9-to-1.0 in 2020, the average yield for the country as a whole rises to 1.72 mt/ha in the terminal year. This 2020 figure is above the 2002-09 average yields for Kazakhstan and Turkmenistan but below all other countries in the region (see figure 3). Area cultivated to wheat in Afghanistan is higher in 2020 than in the base period (see table 3 and figure 10)—returns from planting wheat improve, entirely as a result of the upward trend in yield. Area and yield grow at annual rates of 1.3 percent and 0.9 percent, respectively, and wheat output rises to 4.6 mmt in 2020 (figure 11).

Nevertheless, at 2.2 percent per annum, growth in domestic wheat production does not keep pace with demand, and imports continue to increase (figure 12) 5.2 percent per year. By 2020, approximately 46 percent of Afghanistan's wheat demand is met by imports, versus 38 percent in the base period.

**Table 4. Yield assumptions for Afghan wheat model**

| Variable | Base period | Reference terminal year | High-yield growth terminal year | Self-suffi ciency terminal year |
|---|---|---|---|---|
| | | | *Percent* | |
| Irrigated wheat area | 45 | 45 | 45 | 64 |
| Rainfed wheat area | 55 | 55 | 55 | 36 |
| | | *Metric ton/hectare* | | |
| Average yield on irrigated land | 2.36 | 2.70 | 3.13 | 3.50 |
| Average yield on rainfed land | 0.92 | 0.92 | 0.92 | 1.20 |
| National average yield | 1.57 | 1.72 | 1.91 | 2.68 |

Source: USDA, Economic Research Service.

### *Scenario II: High Yield Growth*

Prices, income, and hence consumption remain the same as in the Reference scenario. Afghanistan's wheat yield increased rapidly from 1990-2009, and Scenario II analyzes the impacts of sustaining this strong growth through 2020. Thus, yield grows at an annual rate of 1.97 percent, from 1.57 mt/ha in the base period to 1.91 mt/ha in the terminal year (see figure 9). Just as in the Reference case, the yield increases obtained in Scenario II do not require that Afghanistan increase the proportion of its irrigated wheat area or that the country improve its yields on rainfed areas. The yield on irrigated fields does however rise from 2.36 mt/ha in 2010 to 3.13 mt/ha in 2020 (see table 4). Note that the yield improvement considered in this scenario is within the genetic potential of existing wheat cultivars discussed earlier in this report.

For the country as a whole, national average yield rises to 1.91 mt/ha in the terminal year. This 2020 figure, which is in line with the 2002-09 average yields for Tajikistan and Russia, is well below those of Pakistan and Uzbekistan (see figure 3).

Area cultivated to wheat now expands more rapidly than in the previous scenario (see table 3 and figure 10), as faster improvements in yield (see figure 9) accelerate the growth in returns from producing wheat. Area, yield, and production increase at annual rates of 1.7 percent, 1.97 percent, and 3.7 percent, respectively. At 5.3 mmt in the terminal year, wheat production in Scenario II is 15.6 percent higher than the 2020 level from Scenario I (4.6 mmt). Nevertheless, with consumption outpacing domestic output, imports continue to trend higher (see figure 12), growing 3.1 percent per annum (see table 3).

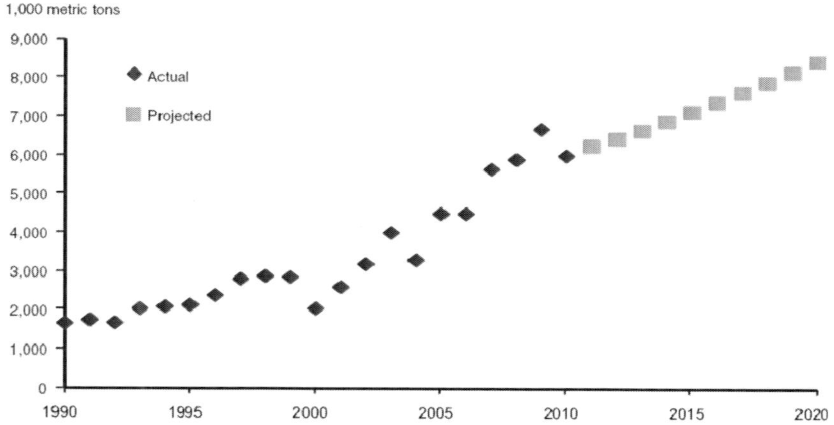


Sources: 1990-2010 data are from USDA, Foreign Agricultural Service, Production, Supply and Distribution (PSD) online database.
ERS model replicates 2010 data; 2011-20 figures are results of ERS model.

Figure 8. Projected wheat consumption (all scenarios).

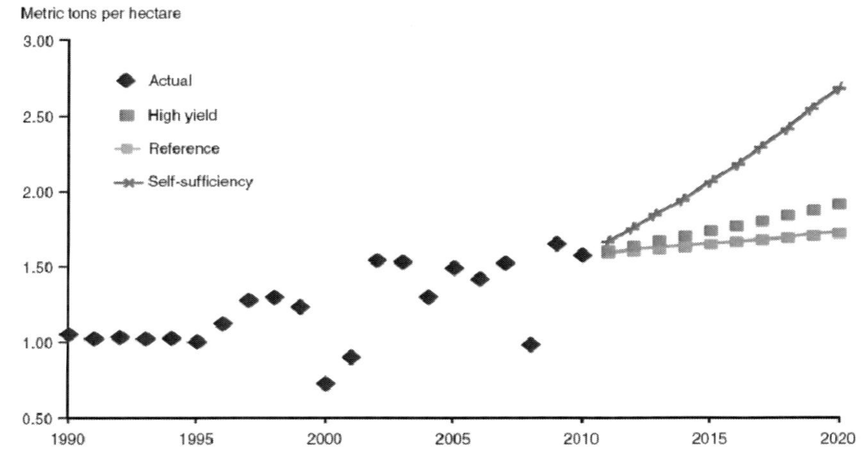


Sources: 1990-2010 data are from USDA, Foreign Agricultural Service, Production, Supply and Distribution (PSD) online database.

ERS model replicates 2010 data; 2011-20 figures are results of ERS model.

Figure 9. Wheat yield under alternative scenarios.

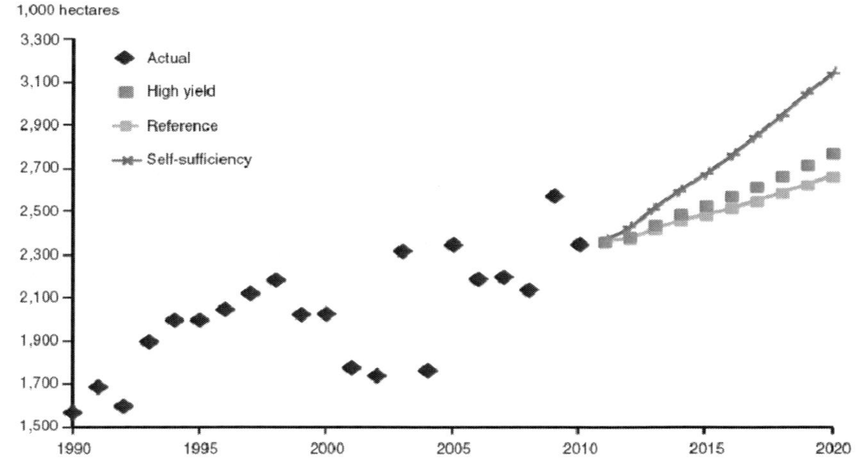


Sources: 1990-2010 data are from USDA, Foreign Agricultural Service, Production, Supply and Distribution (PSD) online database.

ERS model replicates 2010 data; 2011-20 figures are results of ERS model.

Figure 10. Wheat area under alternative scenarios alternative scenarios.

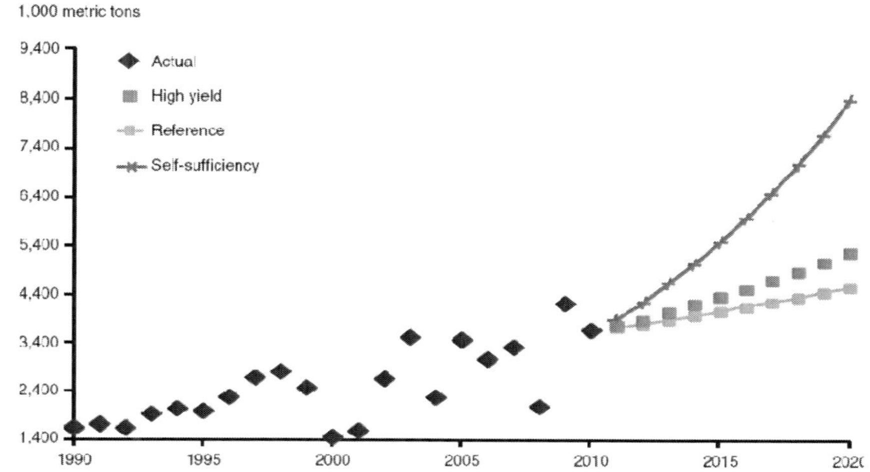


Sources: 1990-2010 data are from USDA, Foreign Agricultural Service, Production, Supply and Distribution (PSD) online database.
ERS model replicates 2010 data; 2011-20 figures are results of ERS model.

Figure 11. Wheat production under alternative scenarios.

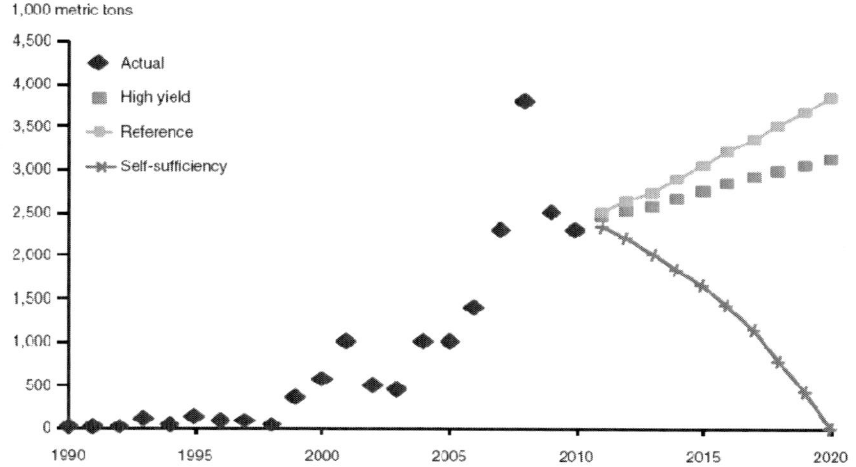


Sources: 1990-2010 data are from USDA, Foreign Agricultural Service, Production, Supply and Distribution (PSD) online database.
ERS model replicates 2010 data; 2011-20 figures are results of ERS model.

Figure 12. Wheat imports under alternative scenarios.

***Scenario III: Self-Sufficiency***

The productivity advances considered in the previous scenario may be somewhat optimistic. Scenario III illustrates, as a hypothetical case, what would be required for Afghanistan to achieve self-sufficiency. Accordingly, this scenario models the impacts of even faster rates of technological change on irrigated fields. Unlike the previous two experiments, yield on rainfed land now grows as a result of adopting higher quality seed. In addition, the proportion of Afghanistan's wheat area that is irrigated now rises as a consequence of improving water availability/management.

Prices, income, and hence consumption remain the same as in the previous two experiments. To achieve self-sufficiency in the next decade, more widespread use of improved seed and fertilizer on irrigated and rain fed wheat fields would have to be combined with improbably sharp increases in irrigated wheat area. In line with the feasible 2020 targets indicated by Maletta (2007), rainfed (irrigated) yields rise to 1.2 mt/ha (3.5 mt/ha), as farmers rapidly adopt improved inputs. The critical factor in Scenario III is that it incorporates unrealistically sharp increases in irrigated area. By the end of the projection period, irrigated fields account for the majority (64 percent) of land cultivated to wheat (see table 4). Consequently, Afghanistan achieves a national average yield of 2.68 mt/ha at the end. This 2020 figure is above the 2002-09 average yield for Pakistan, and is second only to Uzbekistan among neighboring countries (see figure 3). Area cultivated to wheat now expands more rapidly than in the previous scenario (Table 3 and Figure 10), as faster improvements in yield (see figure 9) accelerate the growth in returns from producing wheat. With production increasing at an implausible 8.6 percent per year, Afghanistan closes the domestic supply-demand gap and ceases to be an importer in 2020 (see figure 12). However, substantial growth in Afghanistan's irrigation sector, aside from requiring considerable financial resources, would likely necessitate international agreements with downstream neighboring countries that would face reduced water availability as a result of increased usage in Afghanistan.

## CONCLUSION

In Afghanistan, wheat is a key staple food, accounting for over half of the population's caloric intake on average. Since 1990, Afghanistan has had some success at achieving higher wheat output, driven primarily by yield increases. The historical growth in wheat yields reflects long-term efforts at seed development and increased availability by international organizations. These productivity advances also demonstrate that seed and fertilizer markets have continued to exist and function, albeit with significant disruptions and under a great deal of adversity. This has allowed yield to reach a number of new highs when favorable weather conditions have occurred. However, the historical performance of Afghanistan's wheat production sector also reveals its sensitivity to variations in precipitation. The destruction and continuing disrepair of the country's irrigation systems, aside from exacerbating the impact of adverse growing conditions, is also a likely explanation of the negligible growth in wheat area, which in turn constrained growth in output. Consequently, imports have played a key role in fostering the growth and stability of Afghan wheat consumption.

Despite the country's predominantly mountainous terrain and arid to semi-arid climate, there is long-term potential for increasing domestic wheat production. Irrigation systems have been damaged or have not been maintained, and the country's irrigated area is below its potential. Large areas of Afghanistan are still planted with traditional, low yielding wheat seeds. However, even if Afghanistan sustains its rapid post-1990 production growth, domestic wheat output will not outpace expected increases in consumption, suggesting growing dependence on supplies from Pakistan and other countries. To close the gap between domestic supply and demand, yields on both rainfed and irrigated areas would have to rise. Moreover, self-sufficiency would also entail increases in the proportion of the country's wheat area that is irrigated, i.e., the majority of the country's wheat area would have to be irrigated. Substantial growth in Afghanistan's irrigation sector may necessitate international agreements with downstream neighboring countries, to the extent that they face reduced water availability as a result of increased usage in Afghanistan. Given the difficulties that are associated with expanding the country's irrigation sector, imports of wheat will very likely continue to play an important role in allowing Afghanistan to compensate for its own domestic constraints on water availability.

## REFERENCES

Ahmad, Masood, and Mahwash Wasiq. *Water Resource Development in Northern Afghanistan and Its Implications for Amu Darya Basin*, World Bank Working Paper No. 36, The World Bank, Washington, DC, June 2004.

Asian Development Bank. *Rebuilding Afghanistan's Agriculture Sector*, Manila, Philippines, April 2003.

Chabot, Philippe, and Paul A. Dorosh. "Wheat Markets, Food Aid and Food Security in Afghanistan," Food Policy 32:334–53. 2007.

Dorosh, Paul, and Abdul Salam. *Wheat Markets and Price Stabilization in Pakistan: An Analysis of Policy Options*, Pakistan Institute of Development Economics (PIDE), PIDE Working Papers 2006:5. 2006.

D'Souza, Anna, and Dean Jolliffe. *Rising Food Prices and Coping Strategies: Household-level Evidence from Afghanistan*, World Bank Policy Research Working Paper No. 5466. The World Bank, Washington, DC. November 2010.

Favre, Raphy. *Review of the Wheat Seed Sector in Afghanistan: Implications for Seed Programming and Regulatory Frameworks*, Aizon PLC, Addis Ababa, Ethiopia. March 2004.

Gastel, Antonius G. van, Koffi Amegbeto, and Siddiqi Shamsuddin. *Recounting the impact of emergency seed and fertilizer aids in Afghanistan*, International Center for Agricultural Research in the Dry Areas (ICARDA), Aleppo, Syria, 2007.

Government of Afghanistan. National Rural Vulnerability Assessment Survey. Ministry of Rehabilitation and Rural Development. 2003

Gregory, Ian. *The Role of Input Vouchers in Pro-Poor Growth*, International Center for Soil Fertility and Agricultural Development (IFDC). Abuja, Nigeria. 2006.

Husain, Adil, and Andrew Gilroy. "Afghanistan's Fertilizer Market: Reliant on Imports from Neighboring Countries," Emerging Asia Limited. Shanghai, China. January 19, 2009.

International Monetary Fund. World Economic Outlook Database. IMF, Washington, DC, April 2011.

International Center for Agricultural Research in the Dry Areas (ICARDA), Aleppo, Syria 2007.

International Monetary Fund, *Islamic Republic of Afghanistan: Sixth Review Under the Arrangement Under the Poverty Reduction and Growth Facility*, Country Report No. 10/22. IMF, Washington, DC. January 2010.

Khan, Shamsher. Pakistan *Wheat Subsector and Afghan Food Security: A special report by the Famine Early Warning Systems Network (FEWS NET)*, United States Agency for International Development (USAID). May 2007.

Maletta, Hector. "Food and agriculture in Afghanistan: a Long Term Outlook," in *Afghanistan: How Much of the Past in the New Future*, G.

Picco and A.L. Palmisano (eds). Istituto di Sociologia Internazionale di Gorizia. Gorizia, Italy. 2007.

Maletta, Hector. *The Grain and the Chaff: The Cost of Production of Wheat in Afghanistan in a Farming System Perspective*, Afghanistan Ministry of Agriculture and Animal Husbandry (MAAH) and United Nations Food and Agriculture Organization (FAO), Kabul, Afghanistan. April 2004.

Maletta, Hector, and Raphy Favre. *Agriculture and Food Production in Post-War Afghanistan*, Afghanistan Ministry of Agriculture and Animal Husbandry and United Nations Food and Agriculture Organization. Kabul, Afghanistan. August 2003.

Mollet, Mathias. *After Winter Crop Performance Evaluation of the Emergency Agricultural Input Program–Autumn 2002 Season*, United Nations Food and Agriculture Organization, Kabul, Afghanistan. 2003.

Oushy, Hamdy. Farm Resource Management in Afghanistan–Training Course. United States Agency for International Development (USAID) / New Mexico State University (NMSU) / Afghanistan Water, Agriculture, and Technology Transfer (AWATT) project in Afghanistan. November 2010.

Persaud, Suresh. *Price Volatility in Afghanistan's Wheat Market*, Outlook Report WHS-10d-01, USDA, Economic Research Service. May 2010. Available at: http://www.ers.usda.gov/publications/whs/2010/04apr/whs10d01/.

Rout, Bob. *How the Water Flows: A Typology of Irrigation Systems in Afghanistan*, Afghanistan Research and Evaluation Unit Issue Paper Series. Kabul, Afghanistan. June 2008.

Roy, Amit H. *The Rebuilding of Afghanistan's Agriculture: The IFDC Solution*, International Center for Soil Fertility and Agricultural Development (IFDC), Muscle Shoals, AL. 2003.

Schulte, R.B. *Northern Wheat Trader Survey and Afghan Food Security: A Special Report by the Famine Early Warning Systems Network (FEWS NET)*, United States Agency for International Development. August 2007.

Swedish Committee for Afghanistan. *Agriculture Survey of Afghanistan*, 1992-1993. 17th Report, Islamabad, Pakistan. December 1993.

Torell, Gregory L., and Frank A. Ward. "Improved Water Institutions for Food Security and Rural Livelihoods in Afghanistan's Balkh River Basin," *Water Resources Development*, Vol. 26, No. 4, pp. 613–37, December 2010.

United Nations Food and Agriculture Organization. FAOSTAT database. Rome, Italy. Accessed October 2011.

United Nations Integrated Regional Information Networks (IRIN), Office for the Coordination of Humanitarian Affairs. *Ground water Overuse Could Cause Severe Water Shortage*, Kabul, Afghanistan. September 2008.

United Nations Food and Agriculture Organization. *Review of Global Agricultural Water Use Per Country,* Rome, Italy. 2004.

U.S. Department of Agriculture, Foreign Agricultural Service. USDA Production, Supply and Distribution (PSD) database. Accessed October 2011.

U.S. Department of Agriculture, Foreign Agricultural Service. *Pakistan Allows Wheat Exports*, GAIN Report. December 10, 2010.

U.S. Department of Agriculture, Foreign Agricultural Service. P*akistan Grain and Feed Annual Report 2009*, GAIN Report No. PK9002. March 11, 2009.

U.S. Department of Agriculture, Foreign Agricultural Service. *Afghanistan: Severe Drought Causes Major Decline in 2008/09 Wheat Production*, Commodity Intelligence Report. August 12, 2008.

U.S. Department of Agriculture, Office of the Chief Economist. *USDA Agricultural Projections to 2020*, OCE-2011-1. 2011. Available at: http://www.ers.usda.gov/publications/OCE111/.

# APPENDIX 1

## Elasticities and within-Sample Predictions

Formal estimates of supply and demand elasticities for wheat in Afghanistan are not available in the literature. Limitations in both the quantity and quality of Afghan data, as discussed in Persaud (2010), pose a significant if not insurmountable challenge for econometrically estimating demand and supply parameters.

## Demand Elasticities

Chabot and Dorosh (2007) estimated demand and net imports in 2003-04 under alternative assumptions for production levels and demand parameters. Their three sets of assumed demand elasticities were (1) completely inelastic income and own-price elasticities (2) a low elasticity scenario using income and own-price demand elasticities of 0.2 and -0.2, respectively, and (3) a high elasticity scenario using income and own-price demand elasticities of 0.5 and -0.5, respectively. Chabot and Dorosh's aim was to assess the consistency of reported production and availability changes derived from the alternative data sources with movements in market prices, rather than arguing for a particular set of elasticities.

Recent trends in per capita consumption, the real price of flour, and real GDP per capita are consistent with income and own-price elasticities of 0.32 and -0.2, respectively (close to Chabot and Dorosh's low elasticity scenario) and a 7-percent trend term that reflects the rehabilitation of the country's infrastructure and marketing channels. Thus, a 1-percent increase in real GDP per capita is associated with a 0.32-percent increase in per capita consumption of flour; a 1-percent increase in the real price of flour is associated with a 0.2-percent decrease in per capita flour consumption. These elasticities and the trend term, along with 2000-09 data for real GDP per capita and the real price of flour, are used to compute within sample predictions of per capita wheat consumption.

The assumed demand parameters are plausible in that they generate predictions of consumption that track the actual 2000-09 consumption data reasonably well (appendix figure 1).

It is important to note that the demand projections to 2020 are based on the above-mentioned elasticities while excluding the trend term in order to avoid unrealistically high consumption levels. Rising incomes are expected to lead to dietary diversification in Afghanistan (Maletta, 2007), and current levels of consumption are already high. Projected increases in consumption are expected to be rapid, although not to the same degree as in the past decade.

## Elasticities for Wheat Area Harvested

In the absence of econometric estimates of wheat supply elasticities, Gastel et al. (2007) estimated the "economic surplus" (i.e., producer and consumer surplus) resulting from emergency seed and fertilizer aid based in part on an own-price demand elasticity of -0.1 and assumed own-price supply elasticities of 0.228 and 0.4. Similar to Chabot and Dorosh (2007), Gastel et al. (2007) do not argue for a set of point estimates, and instead use various elasticities to generate scenarios.

Similar to the USDA Baseline models, we posit a relationship between area and expected returns from wheat cultivation, where the lagged wheat price and lagged yield represent expected returns.

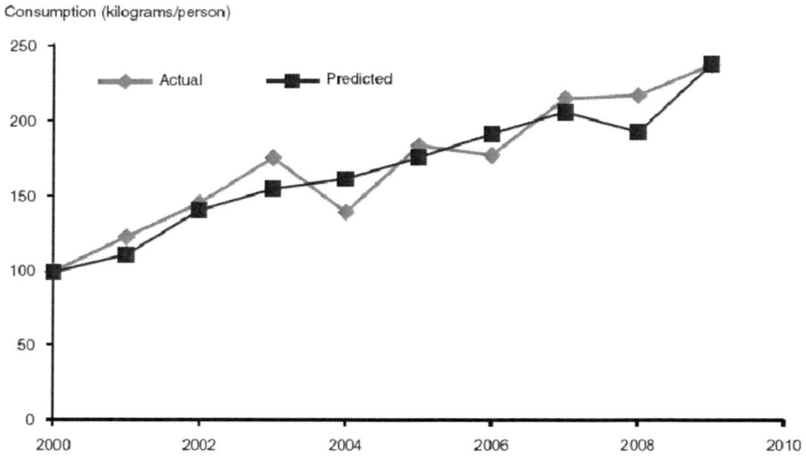


Sources: Actual data are from USDA, Foreign Agricultural Service, Production, Supply and Distribution (PSD) online database and the International Monetary Fund (2010); predicted values are USDA, Economic Research Service calculations.

Appendix Figure 1. Actual and model predictions of per capita consumption of wheat.

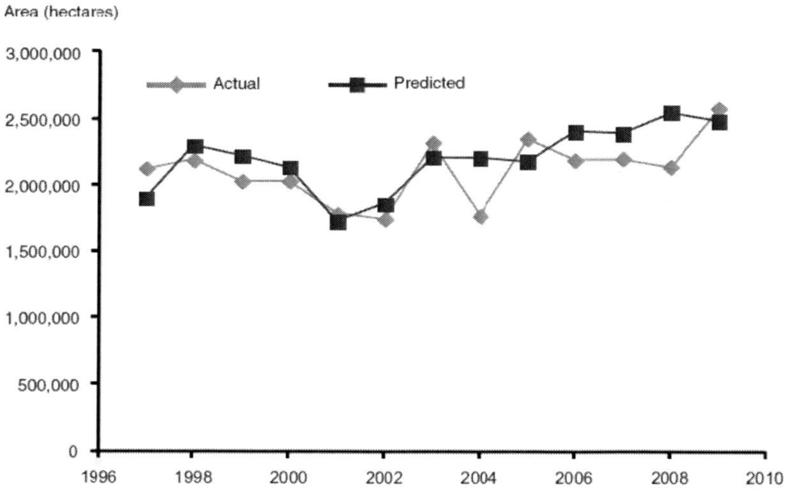


Sources: Actual data are from USDA, Foreign Agricultural Service, Production, Supply and Distribution (PSD) online database and the International Monetary Fund (2010); predicted values are USDA, Economic Research Service calculations.

Appendix Figure 2. Actual and model predictions of wheat area.

It is not possible to obtain data on farm prices in Afghanistan; it is necessary to instead use retail prices of wheat. This approach is reasonable if changes in producer prices mirror changes in available retail prices. Prices of competing crops are not available. However, wheat is the dominant crop in Afghanistan, which would suggest that a given percentage change in the area cultivated to competing crops would have only a small percentage impact on wheat production.[9] On the other hand, since a relatively large amount of Afghanistan's land area is cultivated to wheat, even small percentage changes in wheat area can significantly change the production of competing crops.

Recent trends in area, the lagged real price of wheat, and lagged yield are consistent with own-price and yield elasticities of 0.20 and 0.42, respectively, and a 1-percent trend term. Thus, a 1-percent increase in the real price of wheat is associated with a 0.20-percent increase in wheat area in the following year, all other things remaining the same; a 1-percent increase in wheat yield is associated with a 0.42-percent increase in wheat area in the following year, all other things remaining the same. These elasticities, along with 1997-2009 data for the real price of wheat and wheat yield, are used to compute within sample predictions of wheat area. These model predictions are not capable of capturing or replicating weather-induced contractions in wheat area. However, the assumed supply elasticities are plausible in that they generate predictions of area that, overall, track the growth in the actual 1997-2009 data reasonably well (appendix figure 2).

# APPENDIX 2

## Model Equations

(Variables in bold below are exogenous.)

1. WHPrice = PakRelPrice * (1 + WHMargin)
2. FLPrice = PakRelPrice * (1 + FLMargin)
3. FLDemandPerCap = $FL^1$*GDPPerCap + $FL^2$*FLPrice + constant
4. FLDemand = FLDemandPerCap*POP
5. $WHArea_t$ = $WH^1$*$WHPrice_{t-1}$ + $WH^2$*$WHYield_{t-1}$ + $WH^3$*Trend + constant
6. $WHYield_t$ = $WHYield_{t-1}$*Trend
7. WHProduction = WHArea*WHYield
8. WHImport = FLDemand - WHProduction

## Appendix Table 1. Variable list

| Variable | Description | Unit | Base value |
|---|---|---|---|
| WHPrice | Afghan wheat price, real | US $ / ton | 341 |
| $WHPrice_{t-1}$ | Afghan wheat price, lagged one year | US $ / ton | 318 |
| FLPrice | Afghan flour price, real | US $ / ton | 498 |
| PakRelPrice | Pakistani release price of wheat, real | US $ / ton | 284 |
| WHMargin | Wheat marketing margin | percent | 20.0 |
| FLMargin | Flour marketing margin | percent | 75.1 |
| FLDemandPerCap | Per capita consumption of flour | kgs / person | 199 |
| GDPPerCap | Gross domestic product per capita, real | US $ / person | 517 |
| FLDemand | Total flour consumption | 1,000 tons | 6,000 |
| POP | Population | millions | 30 |
| WHArea | Wheat area harvested | 1,000 hectare | 2,350 |
| WHYield | Wheat yield | tons / hectare | 1.57 |
| $WHYield_{t-1}$ | Wheat yield, lagged one year | tons / hectare | 1.65 |
| WHProduction | Wheat production | 1,000 tons | 3,700 |
| WHImport | Wheat imports | 1,000 tons | 2,300 |

Notes: Base year for deflator set to 2010. The variable for wheat imports (WHImport) includes wheat flour quantities on a wheat-equivalent basis. Flour consumption (FLDemand) is on a wheat-equivalent basis.

Sources: Area, yield, production, and imports are from USDA, Foreign Agricultural Service, Production, Supply and Distribution online database. Afghan prices of wheat and flour are from the U.N. World Food Programme's Price Analysis in Afghanistan (World Food Programme, 2011). GDP deflators, population, and GDP are from the International Monetary Fund (2011). Pakistani release price of wheat is from Agricultural Prices Commission, Government of Pakistan.

## Appendix Table 2. Parameters and elasticities for Afghanistan wheat model

| Equation and coefficient | Elasticity | Parameter |
|---|---|---|
| *Ln(Per capita flour demand) equation:* | | |
| *lnFLPrice (FL2)* | -0.2 | -- |
| *lnGDPPerCap (FL1)* | 0.32 | -- |
| Constant | -- | 4.5129 |
| *Ln(Wheat area harvested) equation:* | | |
| *lnWHPricet-1 (WH1)* | 0.20307 | -- |
| *lnWHYieldt-1 (WH2)* | 0.41714 | -- |
| TREND (WH3) | -- | 0.01 |
| Constant | -- | 9.33874 |

ln = natural logarithm.
FLPrice = Afghan flour price, real.
GDPPerCap = gross domestic product per capita, real.
WHPrice = Afghan wheat price, real.
WHYield = wheat yield.
Source: USDA, Economic Research Service.

# APPENDIX 3

## Data Sources and Uncertainties

There is considerable uncertainty surrounding estimates of population, production, consumption, and trade flows for Afghanistan. Total population estimates are approximations, since large segments of the population flow in and out of the country, including the seasonal movement of nomadic tribes into Pakistan and the large return of refugees in 2002-04, both of which were difficult to count accurately.

The U.N. Food and Agriculture Organization and the U.N. World Food Programme develop crop production estimates for Afghanistan, working under adverse conditions of war, rural insecurity, and poor transport infrastructure, while also coping with tight budget and time constraints. Government data at the provincial level are limited, and this further increases the difficulties of quantifying crop production for Afghanistan as a whole (Chabot and Dorosh, 2007). From 2000 forward, Afghan wheat production estimates from FAO are strongly correlated with USDA Production, Supply and Distribution (PSD) data. This report uses PSD data, since this source provides data through 2009, as well as estimates for 2010.

Estimates of trade flows are uncertain—reliable independent customs data on Afghanistan's wheat imports are not available (Chabot and Dorosh, 2007), and there is a great deal of unofficial cross-border trade (Persaud, 2010). USDA PSD online database may provide the best estimates of wheat and flour imports. Based on USDA's Foreign Agricultural Service (FAS) Global Agriculture Information Network (GAIN) reports as well as conversations with FAS Agricultural Specialists from Islamabad, market sources are utilized to account for official and unofficial trade in the USDA PSD import numbers for wheat and flour. Although trade figures are uncertain, various sources such as FAS GAIN reports, United States Agency for International Development (USAID), the U.N. World Food Programme, and the World Trade Atlas, agree that Pakistan is the dominant supplier of wheat (primarily in the form of flour) to Afghanistan. Kazakhstan ranks a distant second in most years, but revised trade data indicate a significant increase in 2008-09 Kazakh wheat and flour exports to Afghanistan that moderated the drop in Afghan consumption.

This report uses price data collected by the World Food Programme in the six major cities of Afghanistan: Kabul (the capital), Kandahar in the South-west, Hirat in the West, Mazar-e-Sharif in the Northern province of Balkh, Fayzabad in Badakhshan province at the Northeast, and Jalalabad in the

Eastern province of Nangarhar. WFP price data are useful for evaluating the affordability of food for low-income segments of the population. By design, the data series represents the lowest-priced food varieties available in urban bazaars. Consequently, the average retail prices in these cities are likely to be higher than the WFP price series. Nevertheless, the data collection method is straightforward and consistent: WFP data collectors identify the minimum price after recording a sample of prices in different stalls of each bazaar (Maletta, 2004). No data exist on producer prices in Afghanistan.

## End Notes

[1] Farmers also retain wheat for household subsistence.

[2] Per capita consumption is approximated as (total domestic consumption– feed & residual) / population.

[3] In response to the 2007/08 price spikes, D'Souza and Jolliffe (2010) found that ho useholds shifted away from meat, fruits, and vegetables toward staples like wheat as a means of buffering the shock to their caloric (energy) intake. In urban areas, the authors found evidence of Giffen behavior, i.e., increases in demand in the face of price increases.

[4] This figure excludes Afghanistan's opium crop sector.

[5] The prices of food commodities that substitute for wheat may have also exhibited an upward trend, thus discouraging consumers from shifting away from wheat.

[6] Despite being a useful metric, irrigation efficiency, defined as the ratio between the estimated irrigation water requirements and the actual irrigation water withdrawal, does not account for the possibility that unused water can help recharge aquifers (FAO, 2004).

[7] USDA's Baseline includes Afghanistan in the Other Asia & Oceania region.

[8] Personal communication January 9, 2011.

[9] Opium poppy, while a major cash crop, is grown on a relatively small land area, accounting for only about 2 percent of cultivated land in Afghanistan (Chabot and Dorosh, 2007).

In: Global Agriculture ISBN: 978-1-62618-260-8
Editors: M. R. Robertson, J. I. Schmidt 

*Chapter 4*

# ECONOMIC POLICY AND COTTON IN UZBEKISTAN*

***Stephen MacDonald***

## ABSTRACT

Uzbekistan is the seventh largest global cotton producer and third largest cotton supplier for world markets. Uzbekistan's Government policies largely shield cotton producers from world market price signals, and cotton area has changed little over the past decade despite strong international price fluctuations.

Government pricing and exchange rate policies tax cotton producers and more than offset the value of input subsidies for cotton growers. The degree of taxation declined for several years after 2000, but increased again in the late 2000s. In the 2009, cotton output dropped as Uzbekistan responded to reduced water availability and increased global food prices with higher taxes on cotton growers.

With continued taxation of cotton production, Uzbekistan likely will continue to lose ground to more dynamic cotton exporters like India and Brazil. As a result, USDA's longrun baseline projections for Central Asia show that the region's share of world cotton production will continue to fall over the next decade.

* This is an edited, reformatted and augmented version of CWS-12h-01, Publication No. United States Department of Agriculture, dated October 2012.

## INTRODUCTION

Uzbekistan is one of the world's largest cotton exporters.[1] Central Asia[2] accounts for 16 percent of world cotton trade, and Uzbekistan remains the leading producer and exporter among these former members of the Soviet Union. Uzbekistan accounted for 25 percent of world cotton trade in the 1970s and 1980s, more than any other exporter. At that time, its per hectare cotton yield was consistently among the highest in the world, exceeding U.S. yields by 74 percent during the first half of the 1970s (fig. 1).

Since then, Uzbekistan's cotton yield has trended downward, and the Uzbek share of world trade has slipped below 10 percent (fig. 2). Elsewhere around the world, yields have increased, with annual gains among major exporting countries ranging between 1.6 percent (United States) and 4.3 percent (Brazil). Uzbekistan has successfully transitioned into an Asian supplier as its traditional markets in Russia and Ukraine have dwindled, but with output now about half of its 1985 level, it may not be meeting its full potential.

It is questionable whether Uzbekistan could sustain a return to its mid-1980s production levels or whether it would want to. Soviet planners oversaw an enormous increase in Central Asia's irrigated cotton area between 1945 and the 1980s. The consequent streamflow reductions and runoff contamination by fertilizers and other chemicals substantially reduced the area formerly covered by the Aral Sea, resulted in toxic residue along former coastal regions, and reduced crop yields in some provinces. In addition, widespread reports of child labor in Uzbekistan's cotton harvest add yet another dimension to the evaluation of the country's optimal cotton output levels (Environmental Justice Foundation, 2005).[3] Recent weather problems and disputes with neighboring countries over irrigation supplies suggest that, even with a return to mechanization, Uzbekistan's current maximum production potential may be lower than it once was. Regardless of Uzbekistan's potential for sustainable production, a review of the country's economic and agricultural policies shows significant government intervention that currently hinders cotton production, reducing output. This study estimates the degree of taxation implicit in the policies used by Uzbekistan to set the domestic producer price of cotton, and calculates the impact of these policies in terms of a Nominal Rate of Assistance (NRA) (see box, "Nominal Rate of Assistance"). We combine information about Uzbekistan's cotton sector from several recent studies, assess the net degree of direct and indirect taxation of cotton in Uzbekistan, and examine how this taxation has varied over time.

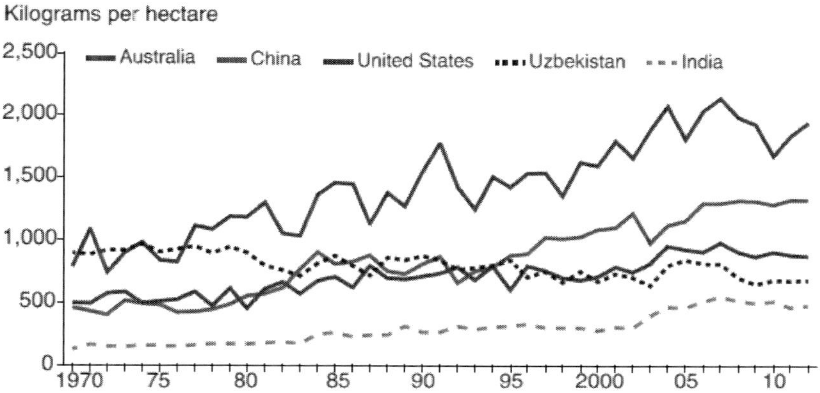


Source: USDA, Foreign Agricultural Service, Production, Supply and Distribution Online data, 2012; USDA, Foreign Agricultural Service, 1977.

Figure 1. Uzbekistan's cotton yields lag.

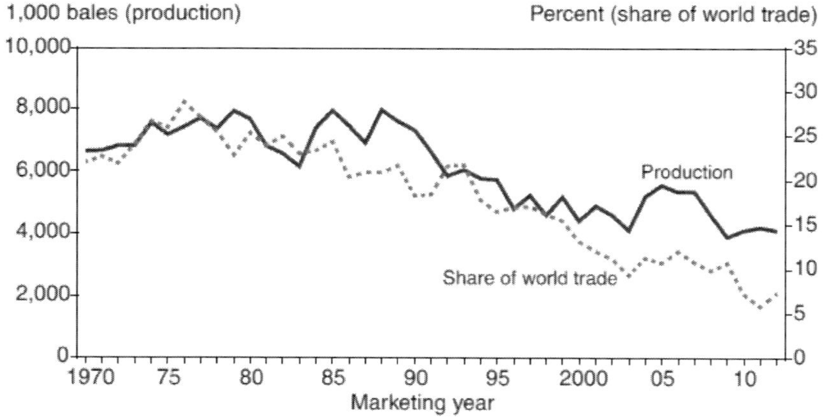


Source: USDA, Foreign Agricultural Service, Production, Supply and Distribution Online data, 2012; USDA, Foreign Agricultural Service, 1977.

Figure 2. Uzbekistan cotton production and share of world trade, 1970-2012.

**Nominal Rate of Assistance**

Measuring agricultural support for a particular commodity requires measuring direct subsidies provided to or taxes imposed upon producers, but also analysis of:

- Relevant trade barriers;
- Support to sectors providing inputs; and
- Whether domestic consumption is subsidized or taxed.

The presence of multiple exchange rates also must be accounted for in countries that enforce capital controls. Josling et al. (2010) summarized the various accounting conventions used to measure agricultural support programs. This study follows the approach described in Anderson et al. (2009) and calculated a nominal rate of assistance (NRA). We used data on the export parity of Uzbekistan's procurement prices—differentiated by official and parallel exchange rates—to estimate the level of implicit and explicit producer taxation (see Appendix for details).

## UZBEKISTAN'S "STATE ORDER" SYSTEM AND COTTON

Since its independence from the Soviet Union in 1991, Uzbekistan has followed a strategy of gradual transition from planned to market economy. The government continues to exercise extensive control in agriculture, particularly in cotton and wheat production, which are referred to as "centralized" crops. The State maintains ownership of the land, and the right to use land for agriculture (other than household plots) is conditional on acceptance of the State's quotas for planting cotton and wheat. The State also provides subsidized inputs, including irrigation. The "State order" system also includes quotas on the production of cotton and wheat, as well as on area planted. To ensure quotas are met, the State monitors efforts year-round (Veldwisch and Spoor, 2008): leaching[4] is monitored in the winter; planting area, varieties, and dates are determined by the State in the spring; and fertilizer application during the growing season is directed by Ministry of Agriculture and Water Resources (MAWR) officials. During the cotton-growing season, State officials visit farms to determine yield potential and adjust planning targets and production quotas. Annual planting area is determined by a State plan and, at the local level, planning may extend to determining which fields are used for cotton, wheat, or noncentralized crops. While the State and collective farms organized during the Soviet period have been largely privatized since 2006, a large number of farms still rely on centralized Machine Tractor Parks (MTP) for machinery, and MTPs prioritize centralized crops.[5] The textile industry exemplifies Uzbekistan's partial transition from planned to market economy. Uzbek cotton consumption has grown in recent years, representing greater foreign investment and an effort by the government to increase the share of fiber processed locally rather than exported. Many textile enterprises are joint ventures, with the government as the main shareholder (Rudenko, 2008). The Ministry of Textiles has been reorganized into the State Joint Stock Company (SJSC) UzbekEngilSanoat, which manages the government's shares in textile firms and is also engaged in promoting the export and domestic sales of local textile output. SJSC UzbekEngilSanoat also promotes investment in the textile industry (Naumov et al., 2010).

## UZBEK COTTON PRICING

The main policy instrument that transfers resources from cotton production is a procurement price set below world prices. Procurement below world prices is possible because producers are effectively required to sell their cotton exclusively through official channels. While much of Uzbekistan's cotton is exported, international trade is strictly controlled. This restriction limits global price transmission and allows resources to be transferred from producers, either directly by the government or indirectly through semi-governmental firms. International cotton prices adjust relatively freely, so exporting countries seldom sustain significant gaps between their export prices and the world price. Uzbekistan regulates its export price, but does so in a relatively transparent manner.

The Uzbek foreign trade companies (FTCs) authorized to export under the Ministry for Foreign Economic Relations, Investments, and Trade[6] price cotton based on the A Index[7] and Cotlook's quote for Uzbek cotton. Exceptions occur in some years, but they are difficult to sustain. For example, Uzbekistan's Government temporarily imposed a minimum export sales price of 60 cents per pound in December 2008, (*Cotton Outlook*, 2009). World prices averaged well below that level for several months during 2008/09, resulting in an unprecedented near-doubling of ending stocks, to 50 percent of use. Policymakers in Uzbekistan have to take the world price of cotton as given when determining export prices, but the determinants of the State procurement price (SPP) are unclear. Djanibekov et al. (2010) suggested that the SPP is established annually based on the world price minus marketing costs. Guadagni et al. (2005) describe such a mechanism as well, but noted that costs were overstated. Sadler (2006) reported that prices were calculated based on production costs adjusted by recent inflation and asserted that, in earlier years, prices were fixed as a percentage of international prices, ranging from 70 percent in 1996 to 85 percent in 1997 and 100 percent in 2000. Rudenko et al. (2009) describe a process where the semi-governmental State joint stock company that monopolizes ginning in Uzbekistan, UzPakhtaSanoat, negotiates a price with the FTCs. For our purposes, this last observation is perhaps the most useful, highlighting the fact that price is determined by balancing competing interests within the Uzbek Government.

Volatile world prices hinder policymakers' ability to precisely target the relationship between the SPP and the world price; they have to choose between recent price movements and longer-term averages when forecasting the world price. One study (Sadler, 2006) noted that, at least prior to 2005, producers entered into a sales contract with their local gin, and that, theoretically, the price would be adjusted if the market price changed. In practice, the price was set so low that even a decline in the market price left it above the contract price. Thus, the system acted as a price smoothing mechanism, reducing SPP volatility compared with the world price (fig. 3). Rudenko et al. (2009) noted that unforeseen price changes between the time the prices are initially negotiated and the time cotton is finally exported are absorbed by the FTCs, effectively smoothing farm prices.

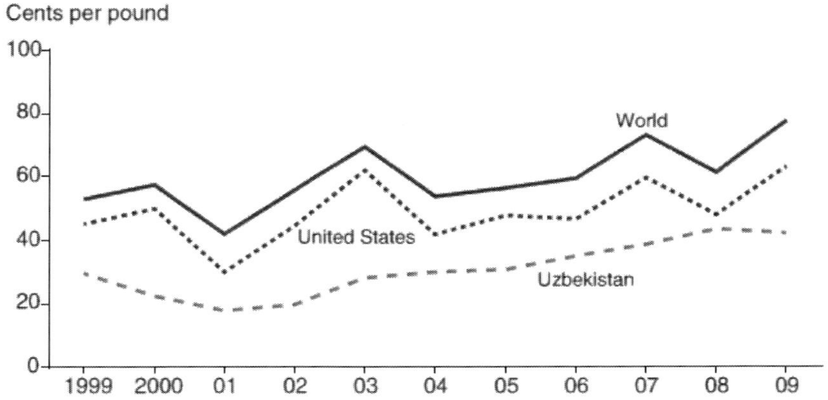


Sources: State procurement prices (SPP) from USDA's attaché in Tashkent and the International Cotton Advisory Committee (ICAC). U.S. cotton price from USDA, National Agricultural Statistics Service, 2011. World cotton price from Cotlook Ltd. Exchange rates from the United Nations Development Programme (2006) and TCX Investment Management Company B.V.

Figure 3. World, United States, and Uzbekistan cotton prices.

Pomfret (2009) also characterized the State procurement system as insulating farmers from world price movements. This characteristic is not necessarily intentional, but is definitely a consequence of policy in Uzbekistan.

## EXCHANGE RATE POLICY

More than 75 percent of Uzbekistan's cotton is exported, so foreign exchange rate policy is an important determinant of the NRA. Uzbekistan's Government controls capital flows and its exchange rate, and cotton and gold are "centralized" exports. Centralization means that all foreign exchange earnings from these commodities must be surrendered to the government for conversion into local currency (the Soum) at the official exchange rate.[8] At times, this policy mix has resulted in a significant implicit tax on the cotton sector as a result of the official exchange rate's overvaluation. An International Monetary Fund (IMF) study of this policy during its peak years (Rosenberg and De Zeeuw, 2001) found that exchange rate overvalution transferred the equivalent of between 3.4 percent (1997) and 6.7 percent (1999) of gross domestic product (GDP) from the cotton sector to the government.

The goals of exchange rate policy can include expansion of trade, controlling inflation, and assistance in achieving longrun development goals. An undervalued exchange rate can promote exports, but an overvalued exchange rate reduces incentives to produce tradable goods. It can even be specifically employed as an export tax.[9] However, an overvalued exchange rate reduces import prices in domestic currency terms, reducing local inflationary pressure. Foreign exchange shortages are likely to result, which can hinder imports but also provide a policy opportunity as sectors can be targeted for assistance through preferential access to the limited supply of foreign exchange.

Uzbekistan's overvaluation first became significant late in 1996 and has fluctuated widely since then. Financial sector reforms starting in 2000 and two large devaluations of the official exchange rate in May 2000 and November 2001 helped bring the official and parallel market exchange rates into alignment by 2003 (table 1). This culminated with Uzbekistan's acceptance of the currency convertibility obligations of Article VIII of the IMF's Articles of Agreement in October 2003.[10] During 2008, however, the official exchange rate again began to diverge from the parallel market rate.[11] Since then, overvaluation ranged from 15 percent in 2008/09 to 48 percent in 2011/12.

When considering the interaction between the level of the SPP and the degree of exchange rate overvaluation, note that cotton policy is probably not the primary determinant of the official exchange rate's level or of its ratio to the parallel rate. Instead, policymakers presumably set the SPP with knowledge of the government's goals with respect to the real exchange rate. While policymakers' targeting of the real exchange rate may not be perfect, the rate has followed relatively clear trends, so the export parity of the nominal SPP can also be targeted to a large extent.

**Table 1. Uzbek cotton procurement prices and exchange rates**

| Marketing year | Procurement (SPP) | Exchange rate (official) | Exchange rate (parallel) | Parallel exchange rate market premium |
|---|---|---|---|---|
| | *Soums per ton* | *Soums per U.S. dollar* | | *Percent* |
| 1999/2000 | 34,800 | 168 | 684 | 308 |
| 2000/01 | 52,000 | 331 | 903 | 173 |
| 2001/02 | 80,000 | 635 | 1,355 | 113 |
| 2002/03 | 126,000 | 918 | 1,153 | 26 |
| 2003/04 | 195,000 | 992 | 999 | 1 |
| 2004/05 | 225,000 | 1,069 | 1,067 | 0 |
| 2005/06 | 255,000 | 1,185 | 1,175 | -1 |
| 2006/07 | 306,000 | 1,244 | 1,268 | 2 |
| 2007/08 | 352,000 | 1,293 | 1,344 | 4 |
| 2008/09 | 430,000 | 1,403 | 1,610 | 15 |
| 2009/10 | 454,850 | 1,537 | 2,101 | 37 |
| 2010/11 | 587,980 | 1,661 | 2,360 | 42 |
| 2011/12 | 496,020 | 1,850 | 2,730 | 48 |

SPP=State procurement price.

Sources: Procurement prices from USDA's attaché in Tashkent and the International Cotton Advisory Committee (ICAC). Exchange rates from the United Nations Development Programme (2006) and TCX Investment Management Company B.V.

## CALCULATING THE EXPORT PARITY OF DOMESTIC PRICES

Accurate information about Uzbekistan's economy is limited, and official statistics are often unreliable.[12] Determining how economic policy affects cotton production is also difficult to calculate due to the pervasiveness of government intervention. The full policy impact on producers is a function of numerous local administrative decisions and likely intentionally lacks transparency. Uzbekistan, however, is not completely closed off from the rest of the world. Procurement prices and exchange rates are available and can be used to examine the relationship between Uzbekistan's SPP and world prices and to estimate whether procurement price policies result in an implicit tax (or subsidy) on cotton production.

This approach is based on a comparison of the ratio of the SPP to world prices against the comparable ratio—prices received relative to world prices—for U.S. farmers. Uzbek farmers, like those in most low-income countries, market their output as unginned seedcotton, so the SPP is a seed-cotton price. In contrast, U.S. farmers gin their own cotton, market the fiber and seeds separately, and receive income from both products. U.S. and Uzbek prices must, therefore, be adjusted to make them more directly comparable. U.S. prices are adjusted by deducting the cost of ginning from the farm price for fiber and then adding in the value of the cottonseed sold separately. The Uzbek SPP for seedcotton is adjusted to a fiber-equivalent basis through division by the proportion of fiber typically derived from Uzbek seedcotton.[13] U.S. and Uzbek cotton are of similar quality, so differences in export parity may indicate whether pricing creates a tax or subsidy for Uzbek farmers.

Pomfret (2009) estimated Uzbek ginning costs for 2002, and Rudenko et al. (2009) estimated the same costs for 2004. Both studies reported ginning costs at around 10 cents per pound, comparable with U.S. costs for 2002 (USDA/ERS, 2012a). Estimated U.S. ginning costs for 1999-2010 are available from ERS's cost and returns estimates; NASS's farm price for cottonseed multiplied by the ratio of U.S. cottonseed to cotton fiber production provides an estimate of the additional value available to U.S. farmers from their farm output (USDA/NASS, 2012; USDA/FAS, 2012).[14] We assumed that transportation costs to export markets were similar, and the distances to Tianjin (China) from both Bandar Abbas (Iran) and Long Beach, CA, suggest this assumption is plausible. Being landlocked—and entirely bounded by neighbors that are themselves landlocked—Uzbekistan faces export impediments that do not affect U.S. exports, but evidence suggests that ocean freight charges are lower for Uzbekistan, which is possibly offsetting.[15]

U.S. policy currently has little effect on the price U.S. cotton farmers receive relative to the world.[16] Therefore, the export parity of U.S. farm prices to world prices should be representative of transportation costs and marketing margins freely determined by market forces. Data from USDA's Farm Service Agency (FSA) support this conjecture. U.S. farm legislation mandates that FSA survey cotton exporters annually to measure average marketing costs of U.S. cotton to foreign markets. During 2004-09, these costs ranged between 19 and 24 percent of the Cotlook A Index (USDA/FSA, 2012). During the same period, the export parity of U.S. farm prices averaged 74 percent, indicating an A Index premium of about 25 percent, consistent with FSA cost estimates (fig. 4). The reliability and speed of delivery of U.S. cotton should allow U.S. cotton to command a premium on world markets, but this premium is largely embodied in the farm prices of U.S. producers.

While adjusted U.S. farm prices for cotton averaged about 25 percent below the world price during 1999/00-2010/11 (table 2), the Uzbek SPP averaged about 50 percent below the world price when valued at official exchange rates and averaged about 60 percent below at parallel market rates, suggesting that the official procurement price substantially underestimates the value of Uzbek cotton on world markets. Within this 12-year period, distinctly different sub-periods can be discerned, with implicit taxes reaching their lowest levels between 2005/06 and 2008/09.

**Table 2. Uzbek procurement and U.S. farm and world cotton prices**

| Marketing year | State procurement price (SPP) Official exchange rate | Parallel exchange rate | U.S. farm price | World price |
|---|---|---|---|---|
| | | *U.S. cents per pound* | | |
| 1999/00 | 29.4 | 7.2 | 45.0 | 52.8 |
| 2000/01 | 22.3 | 8.2 | 49.8 | 57.2 |
| 2001/02 | 17.8 | 8.4 | 29.8 | 41.8 |
| 2002/03 | 19.5 | 15.5 | 44.5 | 55.7 |
| 2003/04 | 27.9 | 27.7 | 61.8 | 69.2 |
| 2004/05 | 29.8 | 29.9 | 41.6 | 53.5 |
| 2005/06 | 30.5 | 30.8 | 47.7 | 56.1 |
| 2006/07 | 34.9 | 34.2 | 46.5 | 59.1 |
| 2007/08 | 38.6 | 37.1 | 59.3 | 72.9 |
| 2008/09 | 43.4 | 37.9 | 47.8 | 61.0 |
| 2009/10 | 42.0 | 30.7 | 62.9 | 77.5 |
| 2010/11 | 50.2 | 35.3 | 81.5 | 165.0 |
| 2011/12 | 38.0 | 25.8 | 90.5 | 103.5 |

Note: Procurement prices for seed cotton converted to fiber-equivalents by dividing by 32 percent ginning outturn ratio.

Sources: Procurement prices from USDA's attaché in Tashkent and the International Cotton Advisory Committee (ICAC). U.S. price from USDA, National Agricultural Statistics Service. World price from the Cotlook A Index. Exchange rates from the United Nations Development Programme (2006) and TCX Investment Management Company B.V.

Between 2005/06 and 2008/09, the SPP averaged only 44 percent below the world price (at market exchange rates).

Using the U.S. average discount from export parity of 25 percent as the free-market cost of transportation and marketing, the additional 19-percentage–point discount relative to the world price received by Uzbek farmers indicated the implicit and explicit taxes resulting from government-administered prices backed by border controls.

Our estimated taxes for 2000/01-2004/05 and the tax changes over time are consistent with findings from earlier studies,[17] but new data indicate that the lower taxes maintained during 2005-08 have recently increased.

## THE ROLE OF MARKETING COSTS

Export parity differences could, under some circumstances, stem from unavoidable differences in marketing costs—such as those introduced by poor infrastructure investment (World Bank, 1999)—but this does not appear to be the case with Uzbekistan.

In developing countries, high costs can be introduced by small-scale marketing systems with poor vertical coordination and integration. Underinvestment in storage facilities can lead to large or potentially large post-harvest losses.

Marketing in Uzbekistan happens on a large scale, with few intermediate steps in the marketing chain and significant vertical coordination. Uzbekistan's relatively arid climate and the seasonality of its rainfall mean that precipitation during key points in processing and shipping is comparable with the U.S. Southwest, where USDA has permitted outdoor storage of cotton placed in the U.S. marketing loan program. Investment in terminal facilities within Uzbekistan—and in railways leading to export points—has been significant as Uzbekistan has adjusted to exporting outside the Commonwealth of Independent States (Cotlook, 2010).

There are inefficiencies embedded in Uzbekistan's marketing system, but they are largely a consequence of government-sanctioned or government-owned monopolies and oligopolies at various stages. Uzbekistan's rigid adherence to export pricing based on widely published world prices provides a useful monitoring mechanism to avoid FTC under-invoicing and off-shore profit capture. Greater returns, however, might be possible with more flexible marketing (Butler, 2005). Similarly, while a greater number of competitors for transportation services for Uzbek exports might eventually lower costs, government policy limits market entry. Such inefficiencies are ultimately implicit taxes on the consumers of these services.

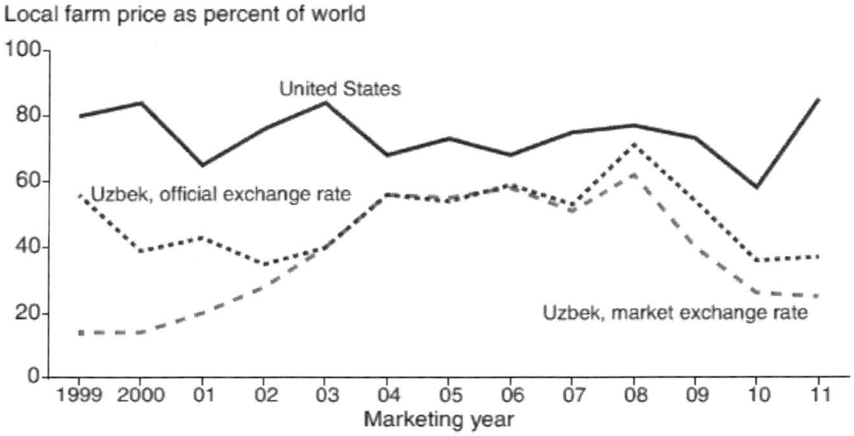


Sources: State procurement prices (SPP) from USDA's attaché in Tashkent and the World Bank. U.S. farm price for cotton and cottonseed from USDA's National Agricultural Statistics Service. U.S. ginning costs from USDA's Economic Research Service cost and returns. U.S. cotton and cottonseed production data from USDA's Foreign Agricultural Service Production, Supply and Distribution Online.

Figure 4. Export parity for U.S. and Uzbek cotton prices.

## NOMINAL RATES OF ASSISTANCE, BY POLICY

The discount from export parity provides an estimate of total implicit and explicit taxes—a negative NRA. This NRA can be disaggregated (table 3) into impacts resulting from SPP policy (NRASPP) and exchange rate policy ($NRA_E$) (see Appendix for details). Both measures are negative in virtually every year, suggesting that both the SPP and the exchange rate policy reduce producer revenues. NRASPP is more stable in certain respects than $NRA_E$ but fluctuates consistent with the price-smoothing characteristic of the policy noted earlier. For example, when world prices fell precipitously in 2008, the NRASPP reached its lowest level of taxation, or 6 percent. Alternatively, when world prices reached record highs in 2010, the NRASPP reached one of its highest levels of taxation, or 52 percent. Taxation through NRASPP trended downward during the early 2000s, stabilizing around 25 percent during 2004-07 before rising during 2010-12. $NRA_E$ remained at or close to zero during 2003-07 but rebounded afterward, reaching 16 percent in 2011.

**Table 3. Implied nominal rates of assistance (NRA), by policy ($NRAS_{PP}$ and $NRA_E$)**

| Marketing year | State procurement price ($NRA_{SPP}$) | Exchange rate overvaluation ($NRA_E$) |
|---|---|---|
| | | *Percent* |
| 1999/00 | -26 | -56 |
| 2000/01 | -48 | -33 |
| 2001/02 | -43 | -30 |
| 2002/03 | -53 | -10 |
| 2003/04 | -46 | 0 |
| 2004/05 | -26 | 0 |
| 2005/06 | -28 | 1 |
| 2006/07 | -21 | -1 |
| 2007/08 | -29 | -3 |
| 2008/09 | -5 | -12 |
| 2009/10 | -28 | -19 |
| 2010/11 | -52 | -14 |
| 2011/12 | -51 | -16 |

Source: USDA, Economic Research Service calculations based on data from USDA's attaché, Cotlook Ltd, United Nations Development Programme (2006), and TCX Investment Management Company B.V.

## SUBSIDIES: EXTENSIVE AND HARD TO MEASURE

Subsidies comprise the other side of the balance sheet in an accounting of the net impact of government policy on cotton, and input subsidies are reportedly extensive in Uzbekistan.

Abdullaev et al. (2009) found that subsidies amounted to $400 million in 2004, or approximately 43 percent of the cotton crop's value, but asserted that they were lower in later years. Pomfret (2009) lists subsidies to cotton producers during 2000-04 that were equivalent to 2.1-8.1 percent of GDP (or $251-$830 million).[18] Guadagni et al. (2005) cited extensive input subsidies ($290-$486 million) during 2000-04 for Uzbekistan cotton farmers that extended to fertilizers, pesticides, and machinery services. Rudenko et al. (2009) noted that farmers received credit at an annual interest rate of 3 percent, compared with 16 percent market rates. Gilham et al. (1995) noted that water was essentially free for farmers. Abdulleav et al. (2009) noted that Water Users Associations (WAU) were established for fee-based water distribution, but that farmers lacked the resources to pay for such services, rendering the payment scheme ineffective.

Veldwisch and Spoor (2008) found that subsidized fertilizer was available to farmers, with the amount distributed determined by their planted area and expected yield. Farmers also were entitled to subsidized diesel, however, the fuel was stored at Machine Tractor Parks (MTP) and distributed only with MTP services.

This study also noted that subsidized farm credit was directed into bank accounts similar to settlement accounts used during the Soviet Era. These same accounts were used to deposit payments for cotton deliveries, while input costs from State-owned firms were deducted directly. Input deliveries may have been partly diverted by officials for personal gain and, according to numerous reports, it is very difficult if not impossible for farmers to withdraw cash from these bank accounts. Rudenko et al. (2009) found that the difficulty accessing the accounts was perceived by farmers as a constraint on cotton production.

Inconsistent subsidy estimates for Uzbek cotton partly reflect the reluctance of officials, traders, and farmers to share valuable information but also reflect the nature of Uzbekistan's economy. Markets for credit and other inputs are poorly developed, and, for example, bartering is commonplace. The conditions placed on the use of subsidized inputs add difficulty to calculating the actual value of input subsidies. On the other hand, without complete markets, assigning a value to State-provided services remains difficult and raises the possibility that their actual value exceeds that assigned in the studies discussed here. When considering the high degree of taxation found in numerous studies, it is important to fully appreciate the uncertainty of these calculations.

Cotton producers do see some benefits from this centralized system. First, producers experience less inherent risk based on the price smoothing aspect of the SPP determination.

Uzbek farmers were protected from shifts in relative world prices during 2000-08 that resulted in a 36-percent decrease in U.S. cotton planted area, compared with a 9-percent decrease in Uzbekistan. Second, the system supports the maintenance and operation of the country's irrigation system. Managing one of the world's more complex irrigation systems is a not an insignificant task, and operation and maintenance issues in Uzbekistan may be less critical than in some neighboring countries.

## Taxation Offsets Subsidies for Uzbek Cotton

The data we analyzed only assessed the implicit taxation of cotton producers through prices. The comprehensive level of net taxation or subsidization depends on the impact of other policies as well. Earlier, comprehensive studies indicated that shifts in net aggregate support for cotton were highly correlated with pricing policy shifts. Shifts in net taxation also were correlated with shifts in Uzbekistan's cotton yields. Recent world price volatility makes analysis of recent pricing policies more complicated but, combined with recent cotton yield data, the shift in the relationship between domestic and world cotton prices suggests that total net taxation increased after 2008. Guadagni et al. (2005) and Pomfret (2009) attempted to assess the overall taxation of Uzbek cotton. Guandagni et al. (2005) completed a detailed study that included input subsidies, exchange rates, the SPP, taxes, and debt restructuring. Debt restructuring was a significant share of government support in some years, which further complicated attempts to calculate support. They determined that, on a net basis, taxation of the cotton sector reached 5 percent of GDP in 2000 and trended downward to 1.8 percent in 2004. As a share of cotton farmers' gross income, taxation trended from 50 to 20 percent over the same period. Pomfret (2009) reported that net transfers from cotton producers as a share of GDP fell from 12.3 percent in 2000 to between 2.1 and 5.4 percent during 2000-04. Results were not entirely consistent across studies. Nevertheless, they show that Uzbekistan's net taxation of cotton declined in 2000-05. Note that the estimated range of net taxation (20-50 percent) underestimates the total negative impact on cotton production in two ways. First, the costs to Uzbek cotton farmers were not fully enumerated even in the studies detailed here. Research by Rudenko (2008) and Velshwisch (2008) found that farmers also face payment delays and skewed classing[19] when marketing their crop. Second, while subsidies partly offset taxes, they do so by introducing an additional set of distortions. Therefore, resource misallocations can persist or can even be amplified. Abdullaev et al. (2009) provided concrete examples of this situation with respect to water management.

## Linking Policies to Cotton Production

Since 2008, increased observed implicit net taxation through prices has coincided with lower yields, similar in both respects to the years prior to 2004.

While producers' efforts and actual yields could vary with producers' expected returns, an additional factor bolsters the link between taxation and reported yields: implementation of the State order system during periods of high taxation likely skews yield estimates for Uzbekistan. Net taxation breaks the alignment of interests between cotton producers and the State. While the State observes producers' efforts to meet quotas, ultimately the asymmetry of information favors producers. To some extent, the asymmetry of information is less with respect to planting activity than with respect to the effort expended to achieve optimal yields. This ensures that some cotton area will be planted that producers later effectively neglect. Data on planting also is likely to be inflated, and the degree of inflation will be correlated with net taxation. Pomfret (2000), USDA's attaché to Uzbekistan (USDA/FAS, 2000), and others have noted occasionally significant smuggling of Uzbek cotton to neighboring countries. Uzbek authorities cannot observe, for example, every time farmers achieve above-average yields, creating opportunities for producers to smuggle cotton out of the country.

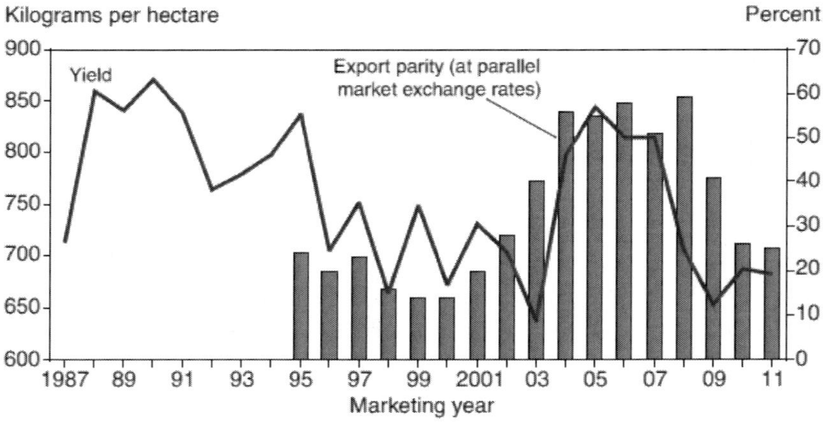


Note: Pre-1995 data is unreliable due to hyperinflation and the difficulties transitioning from Russia's ruble to a local currency.

Sources: Yield data from USDA's Foreign Agricultural Service Production, Supply and Distribution Online and ERS calculations (real SPP) based on data from USDA's attaché in Tashkent and the International Cotton Advisory Committee; the United Nations Development Programme (2006); and TCX Investment Management Company B.V.

Figure 5. Uzbek cotton yield and State procurement price (SPP) export parity.

Thus, while we cannot observe every aspect of net taxation, we can expect reported yields to vary inversely with the level of taxation. Uzbekistan's post-independence cotton yields trended downward but increased during the years when higher export parity indicated reduced taxation through pricing policy (figure 5). Yields jumped to a higher plateau during 2004-07, averaging 19 percent higher than the previous 4 years.[20]

## OTHER INFLUENCES ON YIELD

Other issues may affect cotton yield, but policy differences appear to have accounted for the much of the difference between the two periods (1999- 2003 and 2004-07). Gilham et al. (1995) ascribed the post-independence yield decline to the cumulative impact of limited crop rotation and reductions in chemical and mechanical inputs driven by foreign exchange constraints. The United Nations Environmental Program (UNEP, 2011) highlighted how rising salinity and reduced streamflow have affected yields in Uzbekistan's downstream provinces.

With Uzbek cotton completely irrigated, yield variability is low but not completely eliminated. Significant problems with irrigation supplies in 2000 and 2008 likely accounted for below-average yields. In addition to reduced taxation, good weather in 2004 and 2005 helped sustain yields.

In 2010, growing conditions were very favorable, but yields remained relatively low.[21] Yields in 2010 were even lower than those in 2008, while water intake data from the Amu Darya River indicate more than ample water supplies for irrigation (Scientific Information Center, ICWC, 2012).

Post-2008 data show a rebound in estimated taxation but volatility in world cotton markets must be considered when assessing this data. The price-smoothing aspect of Uzbek policy means that plunging world prices in 2008 resulted in unprecedentedly low taxation levels through SPP (6 percent). Shortly thereafter, record-high 2010 world prices drove the estimated NRASPP to one of the highest taxation levels in this study (52 percent).[22]

Based on the recent shift in cotton yields and a rebound in implicit taxation through prices for almost every year since 2008, we concluded that net cotton taxation has rebounded. Given that yields are currently at their pre-2004 level, net taxation is likely also similar to its pre-2004 level, around 50 percent. Data issues add a degree of uncertainty to our calculations. Subsidy values are difficult to calculate, as are the costs of such administrative procedures as delayed payments to farmers.

Our research substitutes analysis of available aggregate information for a detailed accounting at each step along Uzbekistan's supply chain. But even such an accounting would hold significant ambiguity, given the structure of Uzbekistan's economy.

## FACTORS DRIVING CHANGES IN NET TAXATION

The variations in net taxation of cotton discussed here may be a result of policy shifts in three broader aspects of Uzbekistan's economy: food security, inflation, and the distribution of economic rents.

Negative rates of assistance for cotton production make grain production relatively more attractive to producers and may play a role in food security policy. Uzbekistan's post-independence transition to hard-currency food grain imports was an important factor in the reduction of cotton area during the 1990s.[23] The shift from managed food-grain shipments between republics within the Soviet Union to international market exchange increased Uzbekistan's risk in relying on grain imports. In the mid-1990s, when Uzbekistan began taxing cotton production significantly, its grain import dependency ratio was 80 percent. By 1999/00, it had fallen to 12 percent, where it remained through 2007/08 (USDA/FAS, 2012). The global trade policy environment during this period also supported a reduced emphasis on food self-sufficiency as a result of low price volatility and a strong legacy of increasingly liberal trade arrangements around the world.

In 2008, these trends were disrupted. Irrigation problems throughout Central Asia sharply cut Uzbek grain production, and world grain prices soared sharply. Uzbekistan's main source of imported wheat—neighboring Kazakhstan—briefly imposed an export ban. In response, Uzbekistan announced an area shift for the next season from cotton to grains. The summer of 2008 also marked the reemergence of a parallel market premium for U.S. dollars, reducing the cost of importing grain.

Inflation and food security are closely linked in Uzbekistan, given food's large share of consumer expenditures. The large expenditure share for food and lack of central bank independence makes controlling inflation difficult with volatile food grain prices (Al-Eyd et al., 2012). Lacking efficient markets, there is greater reliance on administrative measures to limit price increases, prolonging the time needed for stabilization. In 2008, Uzbek policymakers explicitly linked exchange rate policy to inflation reduction (IMF, 2008) and, with world grain prices surging once again starting in 2010, this situation likely continued.

Another likely concern for policymakers is the level and distribution of economic rents. Acemoglu and Robinson (2012) characterize Uzbekistan as an economy with extractive economic institutions, and highlight cotton in this respect. IHS Global Insight observed that:

> "Despite a high level of privatization, the country's economy remains tightly controlled by the top circle of leadership. Many, if not most, major private entities are held by members, friends, and family of the ruling elite."

Reduced cotton production has implications for government revenues through excise taxes and to the beneficiaries of government-protected monopolies along the marketing chain. In the 1990s, when Uzbekistan's net cotton taxation rose sharply, cotton accounted for 40-60 percent of export earnings. By 2001, this share had fallen to 25 percent and fell to 11 percent by 2010 as gold and natural gas grew in importance.[24] Increased taxation of cotton production, however, may sustain revenues provided by cotton, even as food security concerns reduce the volume of cotton output and associated rents.

Uzbekistan is often characterized as pursuing import-substitution industrialization (ISI). Under ISI, natural resource-based export sectors are taxed to finance investment in industry. Uzbekistan has had some success with ISI, raising textile output, boosting investment in energy production, and exporting cars. Policymakers may have decided to divert additional resources to industrialization after 2008 but evidence for such a shift appears slim.

Looking forward, two factors likely will determine if Uzbekistan's underlying policy environment continues to support a cotton policy that includes higher taxes on cotton producers:

- Water-related constraints on agricultural production, and
- Continued high world grain prices.

The balance between water availability and needs remains key to both the intensity of Uzbekistan's concern with food security and to local grain production costs. Regional surface temperatures have been trending up, while precipitation has been trending down in the catchment regions of Uzbekistan's irrigation system (UNEP, 2011). These environmental developments, combined with a growing reliance of upstream countries on winter hydroelectric output, have placed greater pressure on summer streamflows (Rakhmatullaev, 2010). As a result, Uzbekistan's food security concerns will remain in the forefront and local grain production costs may also face pressure, suggesting that continued higher levels of cotton taxation could remain to improve the relative attractiveness of grain production.

Trends in world food prices may ease inflationary pressures over time. USDA's baseline forecasts include a reduction in real food grain prices between 2010 and 2020 (USDA/ERS, 2012b). While relatively small, this decline marks a shift from the large increases of 2005-11. Although prices are not expected to return to the record lows seen before 2006, they will help drive consumer prices down. Reduced inflationary pressure could induce Uzbekistan to reduce its exchange rate overvaluation and help moderate the impact of food security concerns on cotton taxation.

## CONCLUSION

Uzbekistan is a landlocked, relatively low-income, and significantly rural nation that inherited a complex economic legacy from decades as a member of the former Soviet Union. Its economic institutions emphasize centralized control of the economy and the appropriation of rents from natural resources for industrialization and for the benefit of the governing elite. Relations with neighboring countries—all former members of the Soviet Union—are complex partly because of shared resources, such as the Amu Darya and Syr Darya rivers. These circumstances are not expected to change in the immediate future, and Uzbekistan's cotton sector likely will grow slowly under a relatively high tax burden.

Shifts in Uzbekistan's cotton policy have exacerbated shifts in world cotton markets since 2000. In the middle of the last decade—when world cotton prices reached record lows relative to other commodities—Uzbekistan reduced the economic burden on its cotton producers, resulting in higher yields, output, and exports and driving world cotton prices even lower. Then, as world cotton markets tightened and prices rose to new peaks, Uzbekistan cut cotton output and exports, partly through increased taxation, adding upward pressure on world prices.

USDA's longrun baseline estimates for Central Asia indicate that cotton production and exports likely will increase slowly over the next 10 years, and the region's share of world output is expected to decrease. The slow pace of economic reform in Uzbekistan, which accounts for two-thirds of Central Asia's cotton output, will remain a key factor driving the modest outlook for cotton in the region. The high level of net taxation illustrated here suggests that significant output gains could be achieved under alternative policies. However, sustaining such gains may require further reforms and investments resulting in more efficient use of inputs, particularly water.

## REFERENCES


Abdullaev, I., C. De Fraiture, M. Giordano, M. Yakubov, and A. Rasulov. "Agricultural Water Use and Trade in Uzbekistan: Situation and Potential Impacts of Market Liberalization," *Water Resources Development* 25(1): pp. 47–63, 2009.

Acemoglu, Daron, and James Robinson. *Why Nations Fail: The Origins of Power, Prosperity, and Poverty*, Random House, 2012.

Anderson, K., M. Kurzweil, W. Martin, D. Sandri, and E. Valenzuela. "Measuring distortions to agricultural incentives, revisited," *World Trade Review* 7: pp. 675-704, 2008.

Al-Eyd, Ali, David Amaglobeli, Bahrom Shukurov, and Mariusz Sumlinski. *Global Food Price Inflation and Policy Responses in Central Asia*, Working Paper No. 12/86, International Monetary Fund, 2012.

Ayers, R.S., and D.W. Westcot. *Water Quality for Agriculture*, FAO Irrigation and Drainage Paper 29 Rev. 1, Food and Agriculture Organization of the United Nations, 1994.

Barros, Geraldo. "Brazil: The Challenges in Becoming an Agricultural Superpower," *Brazil as an Economic Superpower? Understanding Brazil's Changing Role in the Global Economy,* Lael Brainard and Leonardo Martinez-Diaz (Eds.), Brookings Institution Press, 2009.

Butler, R. "Factors Influencing International Cotton Prices," *Cotton Outlook Special Feature: Uzbekistan-2005 International Cotton and Textile Conference,* pp. 6-10, 2005.

Cotlook Ltd. "Uzbekistan," *Cotton Outlook* 88(45): p. 7, 2011.

Cotlook Ltd. *Uzbekistan: Weaving a Way Into the Developed World*, Merseyside, UK, 2010.

Djanibekov, Nodir, Inna Rudenko, John P.A. Lamers, and Ihtiyor Bobojonov. "Pros and Cons of Cotton Production in Uzbekistan," *Food Policy For Developing Countries: The Role Of Government In The Global Food System*, Case Study #7-9, 2010.

Environmental Justice Foundation. *White Gold: The True Cost of Cotton*, London, UK, 2005.

Food and Agricultural Policy Research Institute (FAPRI). *Impacts of Commodity and Conservation Reserve Program Provisions in House and Senate Reconciliation Bills*, FAPRI-UMC Report #15-05, 2005.

Gemayel, E., and D. Grigorian. *How Tight is Too Tight? A Look at Welfare Implications of Distortionary Policies in Uzbekistan*, Working Paper No. 05/239, International Monetary Fund, 2005.

Ghei, Nita, and Stephen B. Kamin. "The use of the parallel market rate as a guide to setting the official exchange rate," Board of Governors of the Federal Reserve System, 1996.

Gillham, F.E., T.M. Bell, T. Arin, G.A. Matthews, C. Le Rumeur, and A.B. Hearn. *Cotton Production Prospects for the Next Decade*, Technical Paper No. 287, The World Bank, 1995.

Guadagni, M., M. Raiser, A. Crole-Rees, and D. Khidirov. *Cotton Taxation in Uzbekistan: Opportunities for Reform*, ECSSD Working Paper No. 41, The World Bank, 2005.

IHS Global Insight. Country & Industry Forecasting, Uzbekistan, Economic Growth: Recent Developments, 2011.

International Cotton Advisory Committee (ICAC). Uzbekistan cotton price database, March 2011.

International Monetary Fund. *Republic of Uzbekistan: 2006 Article IV Consultation—Staff Report; Public Information Notice on the Executive Board Discussion; and Statement by the Executive Director for the Republic of Uzbekistan,* IMF Country Report No. 07/133, 2007.

International Monetary Fund. *Republic of Uzbekistan: 2008 Article IV Consultation—Staff Report; Public Information Notice on the Executive Board Discussion; and Statement by the Executive Director for the Republic of Uzbekistan*, IMF Country Report No. 08/235, 2008.

Isengildina, Olga, C.W. Herndon, and O.A. Cleveland. "Cotton industry in Uzbekistan: Structure and current developments," *Beltwide Cotton Conferences* Vol. 1, pp. 297-301, 1998.

Josling, T., K. Anderson, A. Schmitz, and S. Tangermann. "Understanding International Trade in Agricultural Products: One Hundred Years of Contributions by Agricultural Economists," *American Journal of Agricultural Economics* 92(2): pp. 424-46, 2010.

Lindauer, David. "Parallel, fragmented, or black? Defining market structure in developing economies," *World Development* 17(12): pp. 1871-80, 1989.

Naumov, Y., I. Pugach, and Y. Yusupov. *Uzbekistan's Textile Industry: How to Implement Development Potential?* Policy Brief No. 1(14), United Nations Development Programme, 2010.

Piggott, N.E., and M. Wohlgenant. "Price elasticities, joint products, and international trade," *The Australian Journal of Agricultural and Resource Economics* 46(4): pp. 487-500, 2002.

Pomfret, Richard. "Agrarian Reform in Uzbekistan: Why has the Chinese Model Failed to Deliver?" *Economic Development and Cultural Change* 48(2): pp. 269-84, 2000.

Pomfret, R. "Tajikistan, Turkmenistan, and Uzbekistan," *Distortions to Agricultural Incentives in Europe's Transition Economies,* K. Anderson and J. Swinnen (Eds.), The World Bank, pp. 297-338, 2008.

Rakhmatullaev, S., F. Huneau, P. Le Coustumer, M. Motelica-Heino, and M. Bakiev. "Facts and Perspectives of Water Reservoirs in Central Asia: A Special Focus on Uzbekistan," *Water* Vol. 2: pp. 307-20, 2010.

Rosenberg, C., and M. De Zeeuw. "Welfare Effects of Uzbekistan's Foreign Exchange Regime," *IMF Staff Papers* 48(1): pp. 160-78, 2001.

Rudenko, Inna. *Value Chains for Rural and Regional Development: The Case of Cotton, Wheat, Fruit and Vegetable Value Chains in the Lower Reaches of the Amu Darya River, Uzbekistan*, dissertation thesis, University of Hannover, 2008.

Rudenko, I., J. Lamers, and U. Grote. "Can Uzbek Farmers get More for their Cotton?" *European Journal of Development Research* 21: pp. 283–96, 2009.

Sadler, Marc. "Vertical Coordination in the Cotton Supply Chains in Central Asia," *The Dynamics of Vertical Coordination in Agrifood Chains in Eastern Europe and Central Asia*, Working Paper No. 42, Johan Swinnen (Ed.), The World Bank, pp. 73-114, 2006.

Scientific Information Center, Interstate Coordination Water Commission of Central Asia. Database of the Amudarya river basin, 2012, http://www.cawater-info.net/amudarya/.

Shtaltovna, A., A-K. Hornidge, and P.P. Mollinga. *The reinvention of Agricultural Service Organisations in Uzbekistan – a machine-tractor park in the Khorezm-region*, ZEF Working Paper Series 75, University of Bonn, Center for Development Research, 2011.

State Committee of the Republic of Uzbekistan on statistics. Statistical Review of Uzbekistan, 2011, http://www.stat.uz/en/reports/177/.

TCX Investment Management Company B.V. Personal communication, Jos Kramer, Vice President, 2012.

United Nations Development Programme (UNDP). *How to Boost Exports in Uzbekistan: Policies and Institutions*, Policy Brief No. 4, 2006.

United Nations Environmental Programme (UNEP). *Environment and Security in the Amu Darya Basin*, report prepared on behalf of the partner organizations of the Environment and Security Initiative, 2011.

U.S. Department of Agriculture, Economic Research Service (USDA/ERS). Commodity Costs and Returns: Cotton 2010-11, 2012a, http://www.ers.usda.gov/data-products/commodity

U.S. Department of Agriculture, Economic Research Service. International Baseline Data: Cotton Supply and Use, 2012b, http://www.ers.usda.gov/data-products/international-baseline-data.aspx.

U.S. Department of Agriculture, Farm Service Agency (USDA/FSA). Personal communication, Tom Bickerton, Economist, January 2012.

U.S. Department of Agriculture, Foreign Agricultural Service (USDA/FAS). *U.S. Team reports on Soviet cotton production and trade*, FAS-M-277, 1977.

U.S. Department of Agriculture, Foreign Agricultural Service, Global Agricultural Information Network. "Uzbekistan, Republic of: Cotton and Products Update," GAIN Report Number UZ0012, 2000, http://www.fas.usda.gov/gainfiles/200010/50678605.pdf.

U.S. Department of Agriculture, Foreign Agricultural Service, Global Agricultural Information Network. "Uzbekistan, Republic of: Cotton and Products Annual," GAIN Report Number UZ5004, 2005, http://www.fas.usda.gov/gainfiles/200504/146119488.pdf.

U.S. Department of Agriculture, Foreign Agricultural Service, Global Agricultural Information Network. Uzbekistan Cotton and Products Updates, 1995-2011, 2011, http://gain.fas.usda.gov/Pages/Default.aspx.

U.S. Department of Agriculture, Foreign Agricultural Service. Production, Supply and Distribution Online, 2012, http://www.fas.usda.gov/psdonline/.

U.S. Department of Agriculture, National Agricultural Statistics Service (USDA/NASS). Quick Stats, 2011, http://www.nass.usda.gov/Data_and_Statistics/Quick_Stats/index.asp.

Veldwisch, G. *Cotton, Rice & Water: The Transformation of Agrarian Relations, Irrigation Technology and Water Distribution in Khorezm, Uzbekistan*, Inaugural-Dissertation zur Erlangung der Doktorwürde der Philosophischen Fakultät der Rheinischen Friedrich-WilhelmsUniversität zu Bonn, 2008,

Veldwisch, G.J.A., and M. Spoor. "Contesting rural resources: Emerging 'forms' of agrarian production in Uzbekistan," *Journal of Peasant Studies* 35(3): pp. 424–51, 2008.

Verma, A. "Shipping Cotton to China: Cost and Influencing Factors," *Cotton: Review of the World Situation* 61(5): pp. 16-7, 2008.

World Bank. *India Foodgrain Marketing Policies: Reforming to Meet Food Security Needs, Volume II: Annexes and Statistical Tables*, Report No. 18329- IN, Rural Development Sector Unit, South Asia Region, August 17, 1999.

## APPENDIX: NOMINAL RATE OF ASSISTANCE (NRA)

The discount from the export parity calculated here can be disaggregated into separate impacts resulting from the State procurement price policy ($NRA_{SPP}$) and the exchange rate policy ($NRA_E$) using tools found in the agricultural economics literature. Given that cotton exports are under a complete foreign exchange surrender requirement and that control of domestic and international trade means that domestic consumption of cotton fi ber is based on world price (less a 15 percent discount, which essentially accounts for transportation according to Naumov, et al.), the approach used by Anderson et al. (2009) can be decomposed to illustrate the relative impact of the two policy instruments. The divergence of the State procurement price (SPP) from the world price can be analyzed as an export tax, and the divergence of official exchange rates ($E_O$) from parallel rates ($E_P$) can be treated as a separate additional tax. We define $P_W^*$ as the A Index in U.S. currency terms ($P_W$) multiplied by 0.75 (i.e., 1-0.25, to adjust to a farm-equivalent), then:

$$NRA_{SPP} = \frac{E_P \times P_W^*(1-t_x) - E_P \times P_W^*}{E_P P_W^*}$$

and,

$$t_x = 1 - \frac{\text{SPP}}{P_W^* \times E_O}$$

Following Anderson et al. (2009), when exchange rates are overvalued, the price of exportables is reduced by the fraction $e_x$. Since cotton is a "centralized" commodity with a 100 percent foreign exchange surrender requirement for exporters,

$$e_x = \frac{E_O}{E_P}.^1$$

Thus, the total border support nominal rate of assistance is,

$$NRA = \frac{e_X \times E_P \times P_W^*(1-t_x) - E_P \times P_W^*}{E_P P_W^*}$$
, and

$$NRA_E = NRA - NRA_{SPP}$$

A complete accounting of assistance also would require data on subsidies.

While subsidies for inputs have been signifi cant in some accounting efforts, input market distortions call these measures into question. The cost of late payments and skewed cotton classing also factor into farmers' costs, offsetting some of a given year's input subsidies. As noted previously, earlier studies with access to subsidy estimates from the same period found that taxes on cotton exceeded subsidies, resulting in a net taxation of cotton producers ranging from 20 to 50 percent. The appendix figure illustrates the relationships between world prices *($P_W$)*, the efficient farm equivalent of world prices ($P_W^*$), the SPP, and total taxes and net taxes. The measure of total implicit and explicit taxes, or the NRA, is

$$\frac{\frac{SPP}{E_P} - P_W^*}{P_W^*}$$

NRASPP can be translated into the terms used in the appendix figure and simplified to

$$\frac{\frac{SPP}{E_O} - P_W^*}{P_W^*}$$

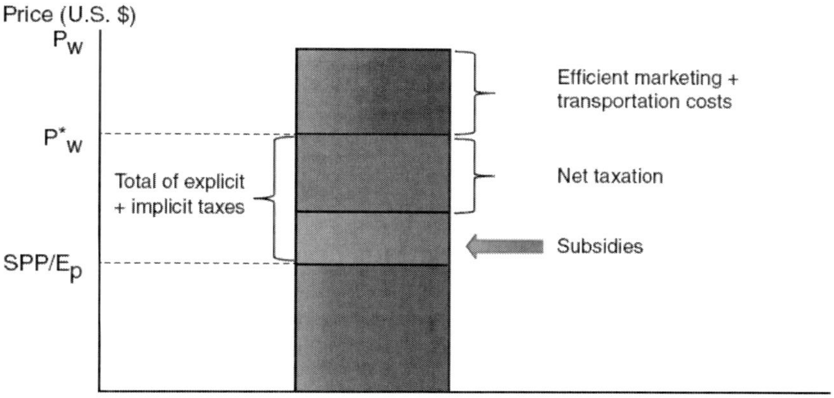


Appendix Figure. Price-based estimate of total and net taxation.

## End Notes

[1] All data on agricultural output, consumption, and trade came from USDA's Foreign Agricultural Service Production, Supply and Distribution (PS&D) database for global agriculture, unless specifically noted otherwise. Units of measurement follow USDA conventions for international data, which include stating cotton data in fiber terms rather than as seedcotton, even in reference to countries where farmers market the latter.

[2] Central Asia refers to the former Soviet Republics of Kazakhstan, Kyrgyzstan, Tajikistan, Turkmenistan, and Uzbekistan.

[3] Since it gained independence, Uzbekistan has largely abandoned the formerly widespread use of mechanical harvesters (Pomfret, 2000), possibly as a result of the disruption of critical supply chains with the breakup of the Soviet Union. Financial distress on Uzbek cotton farms also may have reduced mechanization (Isengildina et al., 1998). The USDA attaché reports that wages for cotton picking are substantially higher in Kazakhstan and the Kyrgyz Republic than in Uzbekistan (USDA/FAS, 2005).

[4] Irrigation can result in soil salinization. Extra water, in addition to crop requirements, can be applied to irrigated fields to leach these salts away (Ayers and Westcot, 1994). In Uzbekistan, this commonly occurs during winter.

[5] MTPs were formerly a component of the State's authority over agricultural producers, but the relationship has evolved into what one study describes as a "patron-client" relationship distinct from formal authority (Shtaltovna et al., 2011).

[6] Three FTCs account for virtually all of Uzbekistan's cotton exports: Joint-Stock Company Uzinterimpex, State Joint Stock Foreign Trade Company Uzmarkazimpex, and Uzprommashimpeks. About 75-80 percent of Uzbekistan's cotton production is exported. Cotton for domestic mill use is purchased through the Uzbekistan Commodity Exchange, which also accounts for a portion of cotton exports.

[7] The A Index is regarded as cotton's world price. Published by Cotlook Ltd, the index is an average of the five lowest quotes for Middling 1 3/32' upland cotton from among 19 types traded internationally. Prices represent offering quotes, for delivery at Far East ports.

[8] Noncentralized exports have a 50 percent surrender requirement.

[9] Brazil's multiple exchange rate system during the 1930s is one example (Barros, 2009). At that time, coffee accounted for 70 percent of Brazil's exports, and coffee export earnings were subject to repatriation at an exchange rate higher than that applied to many other commodity transactions.

[10] However, Pomfret (2009) and others found that access to foreign exchange in bank accounts is problematic, and even access to cash in the form of domestic currency has been difficult (Gemayel and Grigorian, 2005). In addition to the foreign exchange price, the government has administrative policies that can alter the value of foreign exchange holdings, examination of which is beyond the scope of this study.

[11] The unofficial foreign exchange market in Uzbekistan is referred to here as the "parallel" market. Lindauer (1989) assigned the term "black" markets to illegal parallel markets. Press reports from Uzbekistan suggest that parallel markets are currently illegal, but may have been legal previously.

[12] According to IHS Global Insight (2011), "all economic data emanating from the country [are] highly suspect." The IMF regularly notes weaknesses and inconsistencies in Uzbek data that hinder evaluation of economic performance (IMF, 2007).

[13] Piggott and Wohlgenant (2002) described the relationship between the prices of processed goods jointly produced from an input and the input's price. If the proportions of the joint products are a and 1-a; the price of seedcotton is A; the price of cotton seed is B; the cost of ginning is C, and the price of cotton fiber is P, then: $P = [A - (1-a)B + C]/a$. In Uzbekistan, average ginning outturn is 32 percent (a). Here we follow the widely accepted simplification of dividing the seedcotton price by the ginning outturn. This allows direct comparison with the adjusted U.S. price described in the text.

[14] Uzbek farmers are reportedly permitted to purchase byproducts from their ginned cotton at a discount, and Velshwisch (2008) indicated that this occurs at least in Khorezm, but other authors suggest that the practice is not widespread. The discount would likely only add a few cents per pound to the procurement price, changing the calculations minimally. Rudenko (2008) noted that Uzbekistan's crushing industry pays relatively little for the cottonseed it procures and is highly profitable.

[15] Verma (2008) found that transportation from Bandar Abbas to China may be as much as 3 cents per pound less expensive than from the West Coast of the United States, or 4-5 percent of the world price. USDA's attaché indicated that Uzbekistan's supports an excellent transportation infrastructure but, for our purposes, we simplify calculations by assuming that higher domestic costs offset Uzbekistan's external transportation advantage. The need to transfer to a different gauge rail system at the Commonwealth of Independent States (CIS)-Iranian border is an example of a higher cost between farm and port.

[16] User Marketing Certificates (more widely known as the "Step 2" program) were determined by a World Trade Organization (WTO) dispute panel to act partly as an export subsidy. The program was terminated after the 2005/06 marketing year. The Food and Agricultural Policy Research Institute (FAPRI, 2005) estimated that the program added a 2.9 percent premium to the U.S. farm price relative to the world price. This is similar to the 2.3-percentage-point relative decline observed in the U.S. price between 2006/07 and 2009/10 compared with between 1999/00 and 2005/06.

[17] Rudenko et al. (2009) found that Uzbek farmers received 66 percent of the world price in 2004. Guadagni et al. (2005) cited a figure of 77 percent for 2004 and noted that implicit taxes as a result of price controls fell with the unification of exchange rates. Pomfret (2009) cited an

estimate for marketing year 2001 of 54 percent export parity, and presented data consistent with a 37 percent parity using this methodology.

[18]Calculation based on GDP data from IHS Global Insight.

[19] "Classification," for our purposes, describes the application of standardized procedures for measuring the physical attributes of cotton, analogous to grading for grains. Gilham (1995) noted that, for the early 1990s, "Uzbekistan ginneries generally produce a higher percentage of higher grades than the percentage of these grades that are procured," and cited the failure to pay farmers a premium for higher quality as one of the reasons.

[20] Yield variability in each period is low, so the difference is significant at the 1-percent level.

[21] On October 15, 2010, *Cotton Outlook's* Uzbekistan correspondent stated that, "this season's cotton production...has benefited from excellent growing conditions."

[22] Note, however, that the export parity of U.S. cotton shifted significantly that year. Unprecedented price volatility and levels characterized world cotton markets in 2010. Farmers' prices in both the United States and Uzbekistan were largely determined before the mid-season peaks that drove the average 2010 world price to such an unprecedented level. U.S. farmers missed these peaks due to the timing of their marketing, and Uzbek farmers missed these peaks due to the timing of SPP determination and government policy.

[23] Uzbekistan's cotton area fell 29 percent between 1987 and 1995. Cotton area subsequently never exceeded its 1995 level by more than 1.5 percent but trended downward 5-8 percent in 2000 and 2009.

[24] Other energy and oil products accounted for 25 percent; and other major products included food (10 percent), metals (7 percent), machinery (6 percent), and chemicals (5 percent) (State Committee of the Republic of Uzbekistan on Statistics, 2011).

# End Notes for Appendix

[1] Note that the exchange rate in the parallel market, $E_P$, is not the longterm equilibrium rate. The macroeconomic policies associated with the overvaluation of the exchange rate directly and indirectly drive the parallel market rate below the long-term rate, but it is the equilibrium rate in the context of those policies (Ghei and Kamin, 1999).

# INDEX

**D**

## Q



## R



## S